TEXTBOOK OF PETROLOGY

Volume Two

Petrology of the Sedimentary Rocks

Of related interest

TEXTBOOK OF PETROLOGY

Volume One

PETROLOGY OF THE IGNEOUS ROCKS

F. H. Hatch, A. K. Wells and M. K. Wells

Volume Three

PETROLOGY OF METAMORPHIC ROCKS

R. Mason

Petrology of the Sedimentary Rocks

J. T. GREENSMITH
Queen Mary College, University of London

Sixth Edition

London
GEORGE ALLEN & UNWIN / Thomas Murby
Boston   Sydney

First published in 1913
Revised Sixth edition 1978

© George Allen & Unwin Ltd, 1978

ISBN 0 04 552011 9 cased
     0 04 552012 7 paper

British Library Cataloguing in Publication Data
Greensmith, John Trevor
  Textbook of petrology
  Vol. 2: Petrology of the sedimentary rocks—
  6th ed.
  1. Rocks
  I. Title   II. Hatch, Frederick Henry
  552   QE431.2   77–30597

  ISBN 0–04–55e011–9
  ISBN 0–04–552012–7   Pbk.

Printed in Great Britain in 10 on 12 point Times by Unwin Brothers Limited
The Gresham Press
Old Woking, Surrey

Preface to the Sixth Edition

Sedimentary petrology is now an established feature of most undergraduate Geology courses in Colleges and Universities. The courses vary in length, depth of treatment and content, and these days, in view of the substantial perennial increase in data, are almost certainly selective in treatment. The upsurge in process-oriented, quantitative-based research work, bolstered by comparative studies of modern with ancient sediments, is transforming the subject. Plate-tectonic hypotheses have restructured thinking on large-scale sedimentation. But, as stimulating concepts gain ground, so basic ground-work aspects tend to go out of fashion. The petrographic or descriptive side of sedimentary rocks has undoubtedly been downgraded in recent decades, even though it continues to be a part of undergraduate training. The consequences of this neglect are already evident in the research literature. To paraphrase Pettijohn, Potter and Siever (1973) 'the rocks cannot be ignored . . . it is necessary to know how the minerals are put together . . . not to classify or name the rocks, but rather to understand them'.

The petrographic content has not been reduced during the recasting of this book because such material is presumed to be an inviolate feature of sedimentology courses. Some subject matter requiring higher level treatment than befits the objectives of the author has either been expunged or minimised.

The sediments discussed and illustrated are predominantly from the British Isles. This procedure is justified, partly on the basis that common rock types, by definition, are similar wherever they are found and partly because the endeavours of past and present British workers tend to become lost in the whirl of current international research.

In the reshaping of this new edition the author has welcomed the constructive advice of many colleagues, more especially John Allen, Peter Ballance, Tony Dickson, Derek Humphries, Gilbert Kelling, Ken Walton and Brian Waugh. Ed Belt, Graham Evans and John Prentice made valuable comments on the penultimate typescript. The outcome, however, is the sole responsibility of the author. At all stages Roger Jones has been a sympathetic and encouraging listener and advisor. Above all I wish to thank my wife for her constant support and practical assistance in reorganising the contents, and drawing with considerable skill many difficult rock textures, as seen under the microscope, and structures, as seen in the field.

J. T. Greensmith

Contents

Tables

1

Introduction

The fundamental principle upon which the treatment of our subject is based is that of continuity. This principle forms a natural part of the uniformitarian doctrine of Hutton and Lyell, and teaches that the sedimentary portion of the solid crust of the earth was deposited by the ordinary geological agents, namely, gravity, wind, running water, ice, marine action, and so forth, with which we are familiar at the present day. The various processes of compaction, cementation, low temperature metasomatism, and other forms of diagenesis often mask the original character of the older deposits, but comparison with some modern type is generally possible. It is true that there exist among the older rocks types that have been formed by processes not exactly similar to any now in operation, but some analogy can usually be found, and the principle may be said to hold good universally. Many of the older sediments indeed possess characters so similar to those of modern deposits that we can determine with certainty the conditions under which they were formed. Where fossil remains exist such determinations are greatly facilitated.

The word **sediment** in its ordinary interpretation signifies solid material that has settled down from a state of suspension in a liquid. In geological usage, however, it is also employed for many materials which have not settled from suspension in water, such as residual deposits, autochthonous accumulations of organic débris; chemical precipitates, and materials deposited through glacial and aeolian agencies. Many sediments, both modern and ancient, contain a considerable proportion of volcanic ash, which has been transported and deposited by the ordinary agencies affecting clastic material.

A slight acquaintance with the principles of stratigraphical geology suffices to show that under similar conditions similar deposits will be formed. **Similar conditions** must, however, be interpreted as a function of two independent variables. One of these is the actual physical environment of the area where deposition takes place, which may be marine, fresh-water, or terrestrial, and the other is the nature of the material supplied. For example, the deposits arising from a granite as a result of weathering, transportation, and deposition, will be closely similar in different localities, provided that the manner in which these processes operate is identical in each locality; the resulting sediments will, however, be different from those yielded by a limestone in a similar environment. Similar considerations are of great importance in determining the chemical and mineralogical composition of the resulting deposit. The physical characteristics of a sediment, such as texture, structure and organic content, are directly influenced by local conditions, whereas the mineral composition is not only likely to reflect local conditions but also more distant and regional influences.

Under a given set of conditions deposits of a certain general type tend to be formed. The relationship between these and the conditions under which they were formed is summed up in the general term **facies,** this term being used to connote the sum of the lithological and palaeontological characters exhibited by a deposit. Examples of formations presenting the same facies over wide areas are the graptolitic shales of the Lower Palaeozoic systems, the alternating shale–limestone sequences of the Lower Lias, the Greensands of the Cretaceous, the Nummulitic limestone (Eocene), and the modern globigerina ooze. But almost every formation, when traced over a sufficiently wide area, is found to pass laterally into contemporary rocks of a different lithological type. Thus graptolitic shales are represented in other regions by thick beds of sandstone or by limestone, and the Eocene sand and clay beds of Britain are very different from the contemporary limestones of the Mediterranean basin. The marine Devonian rocks and the Old Red Sandstone of the northern hemisphere, though contemporaneous in a broad sense, are entirely dissimilar in facies, and the Chalk of north-western Europe is represented in southern Europe by the Hippurite or Rudistid limestone. Each of the formations just quoted exhibits two very different facies, controlled for the most part by the conditions of formation.

2
Fragmental deposits

Sediments such as gravels, sands, and clays, which consist predominantly of the solid fragments formed by the waste of pre-existing rocks, are grouped together as **fragmental** or **clastic deposits.** Rocks of this group are frequently referred to as **detrital deposits,** a term which, for our purposes, may be regarded as being synonymous with those used above.

CLASSIFICATION

The coarser clastic sediments, such as gravels and conglomerates, consist of rock fragments, each built up of numerous mineral individuals; thus amongst the pebbles of a gravel we may recognise pieces of granite, flint, or quartzite, as the case may be. The medium-textured sediments, the sands and coarse silts, are most commonly (though not invariably) made up of particles, each of which is a fragment of a single crystal, mechanically liberated from the parent rock. The finest-grained clastic rocks, the clays, consist predominantly (but not always exclusively) of minute, flaky crystals formed as the insoluble decomposition products of chemically weathered rocks. A division into these three groups forms a natural basis of classification for the clastic rocks, since it emphasises important differences in the petrological, chemical, and physical properties of these materials. On this basis, then, we shall recognise three major subdivisions of the fragmental deposits:

Rudaceous deposits (e.g. gravels, conglomerates, breccias)
Arenaceous deposits (e.g. sands and sandstones)
Argillaceous deposits (e.g. clays and muds)

GRADE, SORTING AND MECHANICAL COMPOSITION

The term **grade** is used for that fraction of a fragmentary deposit which falls between any two adjacent size-limits; thus we may speak of the silt grade, the coarse-sand grade, or the cobble grade.

The scale of grades commonly used by sedimentologists is shown in

Table 2.1. Certain deposits comprise fragments totally falling within a major size grade, e.g. sandstone and can be named without qualification.

Table 2.1 Size classification and nomenclature of non-carbonate fragmental deposits.

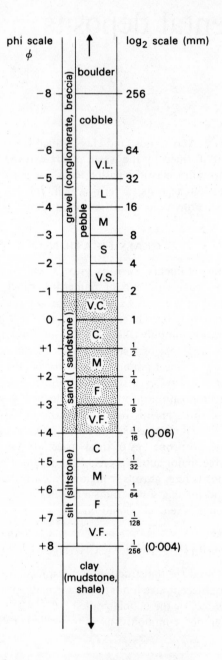

Other sediments have a fragment distribution which extends across major grade boundaries. A boulder clay can consist of large pebble gravel set in a fine silt matrix. Relict gravels are frequently infiltrated by sand and silt. In these examples, descriptive precision is attained by various schemes of which those based on end-member triangles are useful (Fig. 2.1.). Unfortunately,

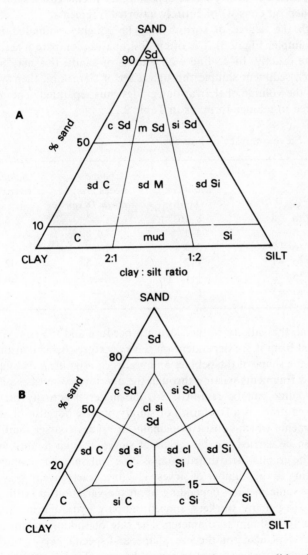

Figure 2.1 End-member textural classifications for unconsolidated sediments based on: A percentage of sand and silt : clay ratio; and B grain size. Each apex represents 100 per cent of each of the three main constituents.

their value is diminished by a lack of agreement on which is the most suitable, so that one scheme may differ considerably from another.

Clastic deposits whose grains are of approximately uniform size are formed under certain special conditions, and are said to be **well sorted** deposits. Most fragmental deposits, however, contain grains of various different sizes, and, according to the degree of admixture of these grades, may be termed **moderately sorted** or **poorly sorted** deposits. In extreme cases, such as boulder clays, the deposit consists of entirely **unsorted** fragments.

Although the degree of sorting may be roughly estimated by the eye (see for example, Fig. 2.3) it is often desirable to compare a series of sediments more exactly. In dealing with gravels or sands, this may be done by putting each sediment sample through a set of sieves and then weighing or measuring the volume of the various grades thus separated. The results may be compared in tabular form, as in Table 2.2.

Table 2.2 Sieve analyses of gravels.

ϕ	Diameter (mm)	Sample 1: beach gravel north Devon	Sample 2: river gravel Glencoe
−3 to −4	8–16	—	6·5
−2 to −3	4–8	4·8	7·8
−1 to −2	2–4	87·7	13·5
0 to −1	1–2	7·5	42·7
+1 to 0	$\frac{1}{2}$–1	—	26·1
+2 to +1	$\frac{1}{4}$–$\frac{1}{2}$	—	3·4

In general, the validity of the sieving procedure and the comparability of data derived from it are dependent on sieve apertures being constant in size. Even then, the shape of the pebbles and grains can significantly affect results, long narrow fragments passing through the apertures whereas equant grains of equal volume can be retained. So, the material actually present on a sieve of given grade is a function not simply of size, but also of shape.

For sediments which consist principally of grades coarser than silt entire analyses can be carried out by means of sieves, though it must be kept in mind that the graph of size distribution is based on only the number of points corresponding to the number of sieves. The finer fractions are best separated in water by some method depending upon the variation in settling velocity with grain size; many workers, indeed, employ hydraulic methods for all material below 1·0 mm in diameter. The rate of sinking of mineral grains in still water depends upon the size, shape, and specific gravity of the grains. Specific gravity is an important factor in the case of sands, and if the sample contains an appreciable proportion of grains with a specific gravity different from that of quartz (2·65), it is preferable to avoid hydraulic methods except for the grades which cannot be separated by sieving. For particles of fine silt, fortunately, the specific gravity has less effect, and in the mud grade there is a limit beyond which particles of the same size and shape are affected similarly in spite of considerable differences in specific gravity. This is due to the insignificance of gravitational differences as compared with the magnitude of surface resistance for very small particles. Again, the shape of the grains is an important factor in determining their behaviour either in still or moving water. A platy grain, such as a mica flake, offers a greater surface area than does a spherical body of equal weight, and consequently is carried into suspension more easily in a rising current, and also sinks more slowly in still water. Many clays consist of minute crystals of flaky and coherent habit, which much increases the surface resistance of the particles in falling through water.

From these considerations it is evident that a mechanical analysis cannot yield results of a high order of accuracy in terms of limiting grain diameters. For this reason, many workers who deal principally with fine-grained sediments or soils prefer to present their results in terms of **settling velocities** or **hydraulic values** instead of in actual terms of particle diameters. For our purpose, however, this would have the serious objection that the coarser grades must of necessity be measured by actual size, since it would be misleading to convert these larger diameters into terms of settling velocities. By making certain assumptions, we can more conveniently convert settling velocities into terms of particle diameter; but it must always be borne in mind that this gives us no more than a conventionalised representation of the facts, and that for the finer grades at least, the diameters calculated from hydraulic values probably depart considerably from the actual dimensions of the particles in question.

In detrital sediments, the grains of sand and silt consist almost entirely of quartz and other minerals of about the same specific gravity. Hence the best procedure is to adopt quartz as the standard, and to fix the relationship between diameter and hydraulic value on the assumption that the material consists of approximately spherical grains with a specific gravity of 2·65. This gives a reasonably near approach to the truth for sands and silts, but the diameters of finer particles, calculated from their hydraulic values cannot be assumed to correspond exactly with the true dimensions of such markedly non-spherical particles. For sediments other than those containing a large clay fraction, it is perhaps best to construct an empirical curve showing the relationship between hydraulic value and diameter for equidimensional grains down to the smallest particles which can be measured by a microscope micrometer. Such a curve, showing the values usually adopted in elutriation, is shown in Figure 2.2.

The component grades of a sediment may be analysed by allowing a suspension to settle in still water, when the larger grains sink rapidly, and the rest of the material subsides more slowly, according to grain size.

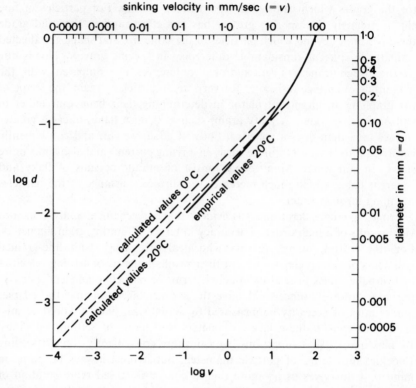

Figure 2.2 Relation between diameter and maximum sinking velocity for particles of quartz settling from suspension in water. The lower part of the curve showing empirical values agrees fairly well with the relationship $v \propto d^2$, but in its upper part steepens until (beyond the limits of the graph) it conforms to the relationship $v \propto \sqrt{d}$.

The most frequently used techniques involve pipettes or sedimentation (settling) tubes. The pipette method is rather crude, but adequate for first approximation analysis. A low concentration sediment suspension is introduced into a graduated cylinder and, at given time intervals, fixed volume samples of the suspension are taken from a fixed level in the cylinder and weighed. A grain size distribution curve can then be calculated from the weight of sediment recovered from the samples.

In the sedimentation tube methods, the rate of accumulation of a known quantity of sediment, introduced into a vertical water-filled tube, is measured by sophisticated balancing machinery. Calculations then enable size distribution curves to be drawn which are more accurate and reproducible than those of the pipette method, and are more amenable to statistical treatment. Moreover, and in contrast to sieving techniques, the curves are based on a continuous sequence of points.

When dealing with highly indurated sandstones and siltstones, sieving and hydraulic techniques are very difficult, so an alternative procedure is to measure grain size in thin-section. The technique consists of measuring the maximum diameter of the clastic grains by using a micrometer ocular. As with all analytical methods, the technique adopted must be standardised. The thin-sections should be uniformly oriented, preferably parallel to the bedding, and individual grains measured at specified intervals along the line of traverse. The results are likely to differ to some extent from the results obtained if the same rocks could be analysed in other ways. Comparison between data derived from using different techniques is rarely feasible and only then with a considerable amount of assumption and mathematical manipulation.

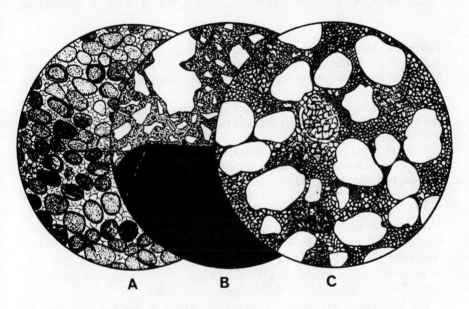

Figure 2.3 Sorting of fragmental deposits.

A, An exceptionally well sorted calcarenite: Aeolian limestone (Pleistocene), Bahamas. The rock consists of well rounded allochems set in a sparite cement.

B, An unsorted deposit: Boulder clay matrix (Pleistocene), northern England. Angular quartz fragments and local rock fragment set in a matrix of fine silt and clay.

C, A moderately well sorted sandstone: Dolomitic sandstone (Ordovician), Virginia.

All the rocks are drawn to the same magnification. The allochems are about 0·3 millimetres in diameter.

Various methods of graphical representation have been devised, one of the simplest being the form known as the **histogram** shown in Figure 2.4.

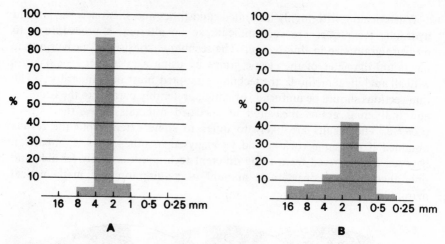

Figure 2.4 Histograms representing the mechanical composition of two detrital sediments. The percentage of each grade is represented by the area of the rectangle. The height of the rectangle is proportional to the percentage which the grade makes of the total sample.

A, Beach gravel, north Devon. This is a well sorted sediment, 87·7 per cent of the fragments having diameters between two and four millimetres (see Table 2.2, sample 1).

B, River gravel, Glencoe. Like most river gravels, this sample shows a poorer degree of sorting than that of beach gravels.

These two histograms are unimodal. Some sediments give bimodal and polymodal histograms as a consequence of even poorer sorting.

The weight of each grade separated is represented by the area of a rectangle. Provided the separations are made at appropriate limits, such that the rectangles may each be drawn of equal width, the area of the rectangle may be measured by its height, and the percentage of each grade present in the sample may then be plotted to a uniform vertical scale.

The method of graphical representation which is used in this book is known as the **cumulative percentage curve** or **summation curve** (Fig. 2.5). The diameter is plotted horizontally, and for any particular value of the diameter, the percentage by weight of particles having diameters equal to or greater than this value is plotted vertically. Thus, if we have a 100 g sample, and 70·5 g are retained on the 1·0 mm sieve (as in sample 2, Table 2.2), we plot an ordinate 70·5 against the diameter 1·0. The sample may then be sieved at other diameters, the total weight held on each sieve being plotted against the diameter of the sieve apertures, until sufficient points have been obtained to draw a smooth curve. It is often convenient to use a logarithmic scale for the grain diameters (as in Figs 2.5 and 2.6) since this allows more accurate comparisons to be made.

It has to be appreciated that there are various scales available which

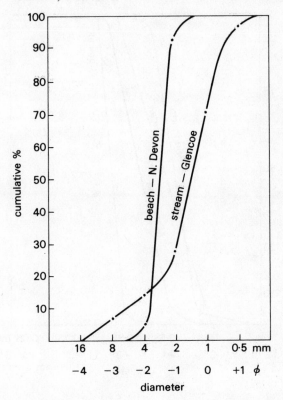

Figure 2.5 Cumulative curves comparing the mechanical composition of two detrital sediments. The well sorted beach gravel gives a steep curve, and the less completely sorted stream gravel gives a less regular and more oblique curve. These curves should be compared with Figure 2.4 and Table 2.2, which represent the same pair of analyses.

express grain size other than in terms of millimetres. One commonly used by sedimentologists is known as the phi (ϕ) scale in which the diameter of grains is recorded on an arithmetic scale calculated from the logarithm to the base 2 of the diameter in mm multiplied by minus one; in other words phi $= -\log_2$ diameter (mm). This looks complicated but, in practice, is relatively simple to use providing conversion tables are available (Table 2.3).

A cumulative curve for a perfectly sorted sediment would appear as a vertical line; where a mixture of grain sizes is represented, the steeper the curve the better the sorting. Mechanically comminuted material, on analysis, is usually found to contain smaller proportions of large and small grains than of medium-sized grains for the sample in question; hence the graph takes the form of a sigmoid curve, steeper in the middle than at the extremes.

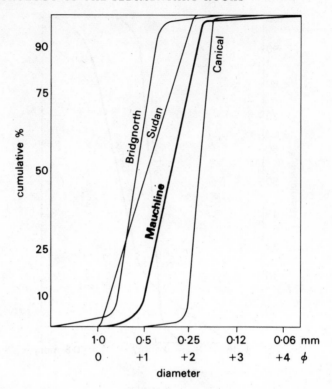

Figure 2.6 Mechanical composition of wind-blown sands. The samples are Bunter Sandstone, Triassic, Bridgenorth, England; Nubian Desert sand, Sudan; Mauchline Sandstone, Permo-Triassic, Scotland; modern basaltic sand, Canical, Madeira. The steepness of the curves indicates excellent sorting.

The degree of sorting is indicated by the steepness of the middle limb of the curve, and by the range of grain size graphically represented by the horizontal distance between the points where the curve cuts the 0 per cent and 100 per cent lines. Wind-blown sands are usually very well sorted, and show extremely steep curves, as in Figure 2.6. Some water-laid deposits, such as the sands of delta-foreset beds and the gravels of sea-beaches, are also quite well sorted. Boulder clays consist of an almost unsorted mixture of rock débris, but if a sample large enough to be representative is analysed, the cumulative curve shows a distinct sigmoid shape (see Fig. 5.5, curves 5 and 6).

Poorly sorted sediments, such as boulder clay, are referred to as well graded by civil engineers, because of the wide range in size of constituent particles. Well sorted sediments, in contrast, are regarded by them as poorly graded.

Table 2.3 Conversion table, millimetres into ϕ units.

ϕ	mm	ϕ	mm
−3·00	8·00	+2·25	0·21
−2·75	6·73	2·50	0·18
−2·50	5·66	2·75	0·15
−2·00	4·00	3·00	0·125
−1·75	3·36	3·25	0·105
−1·50	2·82	3·50	0·088
−1·25	2·38	3·75	0·074
−1·00	2·00	4·00	0·0625
−0·75	1·68	4·25	0·0526
−0·50	1·41	4·50	0·0442
−0·25	1·19	4·75	0·0372
0·00	1·00	5·00	0·0313
+0·25	0·84	5·25	0·0263
0·50	0·71	5·50	0·0221
0·75	0·53	5·75	0·0186
1·00	0·50	6·00	0·0156
1·25	0·42	6·25	0·0131
1·50	0·35	6·50	0·0110
1·75	0·29	6·75	0·0092
2·00	0·25	7·00	0·0078

3

Transportation

It is during transportation that the decomposition products of crystalline rocks undergo a first differentiation, and, as a result of sorting and mechanical modification lose the characters of residual materials, and at the same time a number of incipient sedimentary materials begin to acquire each a new individuality. The soluble salts are carried away in solution, and often travel long distances before they ultimately take part in the formation of chemical deposits or organic limestones. The insoluble products arising from the decomposition of unstable minerals usually consist of fine and flaky crystals; this material, by virtue of its physical condition, is capable of transportation for considerable distances, and goes to form the argillaceous sediments. Finally, the crystals of minerals resistant to weathering are transported more or less unaltered, and go to form the arenaceous rocks.

Sediment is moved by water and air currents and by ice. Although it is outside the scope of this book to deal fully with transportation, some of the salient points can be summarised, more especially those concerning transport by water.

Currents are either undirectional, as in river channels, or multidirectional, as is commonly the situation is shelf seas. In river channels the natural style of water movement is that of unsteady, non-uniform flow. **Froude Numbers** are empirically derived dimensionless parameters based on the depth of flow, mean flow velocity and gravitational forces, and are used to quantitatively define states of flow. Tranquil or subcritical flow, such as might be found immediately downstream of a plunge pool, has numbers less than 1, whereas rapid or supercritical flow, as in a waterfall, has numbers greater than 1. Unsteadiness in flow arises from short and long term variations in discharge and can be quantified by measurement of water velocity at a fixed point. The non-uniformity of flow reflects the magnitude of velocity changes measured along **streamlines.** These are imaginary lines joining a series of particles in the moving water at a given instant of time. The velocity changes are caused by variations in width, depth and slope of the flow. Flow accelerates where channels narrow and decelerates when they widen.

Streamlines in channels are usually complex, diverging and converging or adopting near-circular paths because of irregularities along the bottom. Mobile ripples and dunes are typical examples of bottom irregularity (or hydraulic roughness) which affect uniformity of flow and thus control the amount of sediment being held in suspension. Bottom irregularities also help to create turbulence, which can be considered as random eddies super-imposed onto the main flow. The eddies are generated by relatively intense shear forces, as when a current flows (or jets) across the crest and steep lee slope of an asymmetric ripple or dune. They are an important factor in holding material in temporary suspension.

Turbulence is often expressed in terms of another empirically derived dimensionless parameter, the **Reynolds Number,** which is based on the ratio between inertial and viscous forces. At low numbers, less than 1000–2000, flow tends to be laminar with adjacent flow layers moving by each other without much mixing. Silt and sand grains are gently rolled along the bottom and very small ripples can be produced. At higher numbers, turbulent flow and mixing of flow layers are usual and bring about saltatory movements and intermittent longer periods of suspension of particles. Silt and clay grains are readily held in suspension.

The competence of currents to move grains is affected considerably by increased turbulence and bottom irregularities, such that the shearing stresses along the bottom, which mobilise grains, may be raised in value by three or four times. This may be important in lifting fine silt and clay grade materials from the bottom. These materials are more difficult to move than sand grains, because they compact into cohesive aggregates more readily and tend to firmly adhere to the bottom. Comparatively high shearing stresses are required to lift mud, though the corrasive effects of coarser sediment already held in suspension probably diminish the difficulty of lifting.

Once mud is in suspension, with other grades of material, the particles as a whole tend to be hydraulically sorted on the bases of size, shape and density. This process can lead to the formation of bodies such as pebble lag-concentrates and placers.

On intertidal flats bounding estuary channels, there is a general decrease in grain size of material from the channels across the flats to the marshes. The progressively decreasing velocity of the currents leads to a reduction in the capacity and competency of the waters so that the coarsest material is deposited first and the finest material last. When the deposition of the finest materials in the innermost zones close to high water mark is dominant then a progressive build-up and seaward extension of the salt marsh and high intertidal flat zones occurs. Such is the situation in The Wash of eastern England. 'Settling lag' appears to have a bearing on the effectiveness of sorting processes and salt marsh extension. This term refers to the fact that there is a time lag between the moment at which a current of decreasing velocity can no longer hold a particle in suspension and the moment at which the particle reaches the bottom. The finer the particle the greater the settling lag, hence the ordinary sorting processes are emphasised and the probability of the finest material being deposited at high-water mark at the expense of the coarser is increased.

Estuaries are limited to river mouths in tidal seas. Estuary waters have varying degrees of salinity depending on the relative amounts and degree of mixing of fresh and salt water which is introduced. In many estuaries there is a tendency for saline waters carrying bottom suspended sediment to move upstream along the bottom during rising tides like a wedge, then return seawards during ebb tides. A surface wedge of fresh water moves in a complementary fashion, a feature particularly well demonstrated at the Mississippi mouths where river flow is dominant. In estuaries where tidal currents predominate the mixing of fresh and salt water is much greater, which induces a considerable degree of homogeneity in chemical and physical qualities. The state of the water has a bearing on the vertical and lateral distribution of organic and inorganic materials in transport and in process of being deposited. It is hardly surprising, considering the instability of environmental factors, that there are many variations in the distribution of sediments in estuaries. The effect of conflicting currents upon deposition is especially seen at the mouths of estuaries, where the growth and demise of off-shore banks, and migration of associated channels, take place. The mobility of these superficial structures, in turn, affects the local current patterns, biologic distribution paths and, ultimately, the kind of sediment deposited at any given point.

Most particles suspended in estuaries are finer than 100 microns and a high proportion consist of organic matter. When this finely divided matter comes into contact with a saline solution a physical change in its condition is brought about, the fine particles coagulating to form a flocculent precipitate, which in still or gently moving water falls to the bottom. This process occurs at or near the mouths of rivers where the fresh water first mixes with the salt sea-water, and this fact accounts in part for the prevalence of mud in and adjacent to estuaries.

One of the results of flocculation is the formation of dense suspended pockets or clouds of turbid water which move up and down the deeper channels of estuaries in response to rises and falls in tides. A dynamic balance seems to be maintained within these pockets in that the material which is deposited from them, usually rather quickly at slack water, is replaced at a similar rate by freshly flocculated material. The sedimented mud is very mobile and capable of redistribution by flow down suitable bottom slopes.

The measurable difference between the current velocity at which particular particles are deposited and the greater velocity at which they can be set in renewed motion is known as 'scour lag'. The additional energy required to lift the particles against the forces of adhesion and gravity is greatest for clay-grade particles. On intertidal flats the effects of partial exposure and dewatering of the sediment, especially if fine grained, usually mean that the scour lag is even greater and comparatively high current velocities are needed before the material can be put into motion once more. The agglomerating activities of intertidal organisms may also stabilise mud and influence attempts at remobilisation by currents.

Estuaries and tidal flats are restricted areas of fine sediment transport and accumulation. Near the edge of continental margins, and over the floor of the Atlantic and Polar Oceans, a relatively permanent suspension of fine material (particles usually less than twelve microns) about one kilometre thick has been observed. This is the **nepheloid layer.** The particles are terrigenous clay predominantly and have been derived mainly from shelf areas. They are held in suspension by turbulence induced by surface storm waves, inter-reactions at the interface between powerful surface and bottom currents, and the movements of occasional turbidity currents. Although constant turbulence keeps most particles in suspension, some eventually escape and settle to the sea floor.

Two other styles of transportation in deep waters are where sediment is moved along the bathymetric contours of submarine slopes by deep currents and where sediment is moved by turbidity currents.

Deep sea currents have measured velocities as high as 70 cm/s and are capable of eroding and moving coarser mud, silt and sand particles; small pebbles are moved less frequently. The resultant deposits, known as **contourites,** are usually thin and well laminated with parallel and cross-laminated structure. Less often, towards the base of the continental rise and on adjacent abyssal plains, the mobilised sediment is moulded into asymmetric sandwaves with amplitudes of several tens of metres and wavelengths of a few kilometres.

Turbidity currents are unsteady gravity flows of suspended sediment, whose movement depends on an excess of specific weight over the surroundings. They usually have very large Reynolds Numbers and almost certainly are in a supercritical state.

The causes of the initiation of turbidity currents are not fully understood. Sediment instability on bottom slopes as low as a few degrees may be a function of high rates of deposition leading to underconsolidation of the material. This means that the material is laid down so quickly that interstitial water is trapped in quantity and excessive pore pressure is developed. Hence, the sediment has a lower shear resistance than when normally consolidated and is likely to slump or slide by 'spontaneous liquifaction' when abnormally loaded. Earth tremors of great magnitude may be an important contributory cause of failure. As the material slumps, water may be gradually taken up, so reducing viscosity and increasing velocity. At this stage, when the water content is still not too high, the moving sediment may be in a 'fluxo-turbidite' state with little turbulent mixing of the particles.

If acceleration continues there is further admixing of sediment and water, especially at the head of the flow, and the density decreases and turbulence increases; a turbidity current has now formed. The coarser material is concentrated towards the head and bottom of the current, hence the specific weight of these parts is higher. This results in the coarser material in the lower head region of the flow moving relatively faster than the material in the upper rear parts. Velocities at the flow head are likely to be of the order of one to several metres per second.

The variations in velocity and sediment dispersal within the turbulent head of flows are considerable. Erosional sole structures, such as flute marks, are initiated here. Flute marks are generally arranged in groups parallel to the flow direction and reflect persistent scour of the underlying mud through the full period of current movement (Fig. 4.9). Tool marks, in contrast, are less uniformly oriented and indicate impact and erosion of the mud by some moving object being carried by the vortices within the current. Tool marks often criss-cross each other and associated flute marks at angles of up to thirty degrees.

The slowing of turbidity currents brings about a diminution in their thickness, density and turbulence, so that deposition with vertical grading ensues (Fig. 4.8). Lamination and current ripples form as velocities fall and grain size decreases. Ultimately, the finest particles from the upper and tail parts of the current are deposited.

4

Sedimentary Structures

Structures formed entirely by sedimentary processes are difficult to classify satisfactorily and most attempts have resulted in mixed morphologic–genetic groupings. Dunes have a characteristic shape (or bed form) created by movement of loose particles. They can be regarded as representative of many structures formed by external (**exogenetic**) forces. In contrast, structures such as load casts, generated within sediments after deposition, but prior to lithification, can be regarded as **endogenetic.** The contemporary activities of organisms produce **biogenic** structures.

RIPPLES, DUNES AND CROSS-BEDDING

Progress has been made in understanding how subaqueous ripples and dunes form by studying flow in alluvial channels and flumes, whose beds are composed of non-cohesive materials. Within these channels the flow conditions, such as current velocity, resistance to flow, shear stress at the bed, viscosity and the effect of gravity can be approximately determined and assessed in terms of flow regime. **Flow regimes** can be defined as a range of flow conditions having similar sediment transport and resistance characteristics that produce similar bed forms. Under lower flow regime conditions loose sand is moulded into plane beds, ripples, megaripples, ripples on dunes, dunes and sandwaves (Fig. 4.1). These structures reflect progressively increasing current velocities. The sediment particles appear to move intermittently, rolling up the gentle upstream stoss-slope of the structures and sliding or avalanching down the steeper downstream lee-slope. The upper flow regime is typified by plane beds and antidunes. In general the particles appear to move much more continuously than in the lower regime.

Plane beds formed under high current velocity conditions have no surface irregularities larger in amplitude than a few grain diameters. On the upper surface of flat laminae in many ancient fluvial and littoral sandstones high velocities are expressed by very faint *en-echelon* ridge and hollow structures, known as **parting** or **primary current lineations.** The long axes of the con-

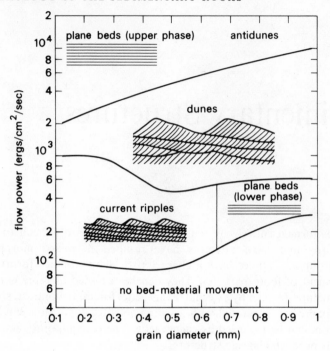

Figure 4.1   Bed forms, flow power and grain size.

stituent grains roughly parallel the elongation of the structures and the largest grains are invariably concentrated in the ridges.

Also at the highest velocities dune-like structures take on regressive migratory qualities, actually moving upstream against the current flow; they are then called **antidunes.** Although observed with amplitudes of as much as one metre in braided channels, antidunes tend to be destroyed as current velocities diminish.

Ripple marks are environmentally unrestricted and are formed by wind activity as much as by water movement. As with all bed forms they can be composed of particles other than quartz sand and silt. They are commonly small-scale, asymmetrical or symmetrical, with successive crests less than 30 cm apart and with amplitudes less than 30 cm. Wind ripples are usually flatter and less symmetrical than current ripples; their crests are straight or gently curved.

The trend of the crests of ripples depends on the trend of the water or air currents in any particular environment. In shallow water areas the crests of small-scale ripples generally subparallel the nearby strandline. This is particularly well seen in sandy and silty areas of deposition, such as bays and intertidal flats. Away from the influence of strandlines, ripple crest trends are more random.

The characteristic internal structure of most ripples is cross-lamination developed during down-current movements. Sand is eroded from the gentle up-current side and mostly deposited towards the top of the steep lee-slope. At intervals this sand avalanches down the lee-slope, in doing so producing the laminated, forward-growing structure.

In some situations the ripples seem to climb on to the gentler stoss-slope of the ripple immediately down-current. The angle of climb is commonly between 5 and 20 degrees from the horizontal. The term describing this arrangement is **ripple-drift cross-lamination** and beds composed of piled-up sets of this type, each set a few centimetres thick, are common in the sediments of fluvio-deltaic successions and some turbidite basinal successions (Fig. 4.2.). They are also described from glacial outwash deposits.

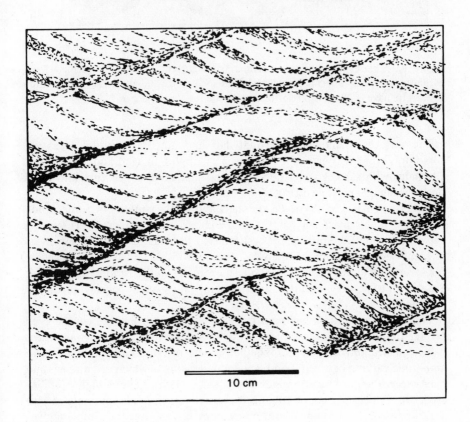

10 cm

Figure 4.2 Ripple-drift cross-lamination. Forward movement of the successive ripples has been from left to right. This common variety of the structure has erosional contacts.

Figure 4.3 Flaser bedding: A, formed from current ripples; B, formed from oscillation ripples. The mud (black) lenses in A and sand-silt lenses in B often appear to have a 'floating' relationship to the rest of the deposit.

The surfaces of modern intertidal flats and adjacent subtidal areas are commonly rippled. Where the formation of ripples on sand and silt surfaces alternates with slacker water deposition of finer mud, a structure known as **flaser bedding** is produced (Fig. 4.3).

In certain major sandy river channels and at the mouths of some sandy estuaries ripples reach an amplitude (height from crest to trough) varying between 30 cm and 2 m and are best referred to as **megaripples.** Their crests in plan view are straight, linguoid or lunate in form and may be traceable for several hundreds of metres. Modern examples in the Brahmaputra River are known to be very mobile and movements up to 200 m per day have been recorded.

The size of structures having a basic ripple-like morphology can be even greater than that of megaripples and at this point different terms are used, namely **dunes** and **sandwaves.** Whether water- or wind-laid dunes have an amplitude varying between 2 and 8 m and wavelengths which often reach and may exceed 500 m. Again, water deposited varieties have a high mobility especially during phases of enhanced current velocity. The ultimate in size are sandwaves well known on many present-day epicontinental shelves where they can be readily identified using bottom sounding apparatus. Amplitudes commonly reach 15 m and wavelengths may be as little as 200 m or as much as 1000 m. The crests tend to be straight and sometimes can be traced laterally for several hundreds of metres. Megaripples may be super-imposed.

The movement of particles in migratory dunes and sandwaves is similar to that of ripples, but creates a much thicker type of internal structure referred to as **cross-bedding**. Cross-bedded sets can be five or more metres thick. The thickest are most often found in wind-blown deposits, whereas the sets in shallow water deposits (gravels, sands, calcarenites) are generally less than three metres. Several varieties are recognised (Fig. 4.4) such as **tabular** with relatively planar surfaces bounding the unit (or set) and **trough** with curved surfaces beneath the unit. Sometimes these bounding surfaces in either type are non-erosional but, more often, are erosional in origin. All varieties of cross-bedding tend to grade into each other and can be seen closely intermingled in some thick sandstones which comprise several units or sets.

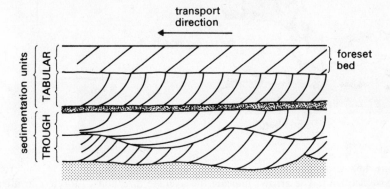

Figure 4.4 Cross-bedding terminology.

Some types of cross-bedding can be useful as an indicator of current and sediment transportation directions. In practice, as many foreset directions as possible are measured within a given bed or group of closely related beds and, after making corrections for tectonic dip, the data are plotted on a suitable base map. Figure 4.5 illustrates how tabular and trough cross-beds are plotted for the Calciferous Sandstone Measures of eastern Scotland and the statistical treatment adopted to emphasise the dominant flow and transportation directions.

It is generally believed by most workers that the distribution of cross-bedding directions within sandstone beds can indicate the source direction and regional slope during the period of deposition (Fig. 4.6). This assumption is probably valid in fluvial and deltaic sequences where the dominant foreset direction may be taken as pointing away from the source area, but it has to be realised that marine transporting currents often flow in a long-shore fashion parallel to nearby coastlines.

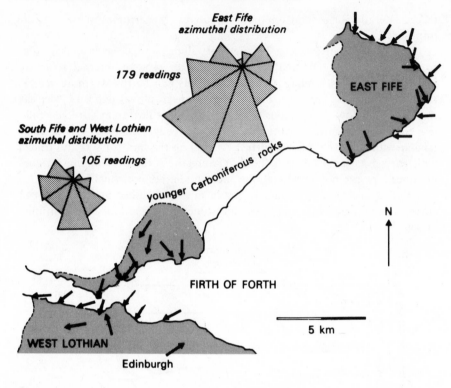

Figure 4.5 Cross-bedding in the Calciferous Sandstone Measures (Carboniferous) of east Scotland. The arrows indicate the average direction of foresets in fluvio-deltaic sandstones at a particular locality. The azimuthal distribution of foreset directions in the two, now disconnected, areas indicates a predominantly southerly and south-westerly movement of sand into the basin of accumulation.

The cross-lamination of clayey silts (lateral deposits) found in intertidal flat successions, and caused by a point-bar style of deposition on the meander core of migrating creeks, is also unlikely to prove helpful in determining regional slope and provenance (Fig. 4.7).

GRADED BEDS AND SOLE STRUCTURES

Fining-upwards graded beds, such as occur in thick flysch successions, are deposited from turbidity currents and are referred to as **turbidites** (Fig. 4.8). The beds are usually and predominantly of graywacke lithology but, occasionally, are clastic limestones.

Linear (sole) structures occur at the base of many turbidite units and the majority are caused by the scouring action of currents from which the

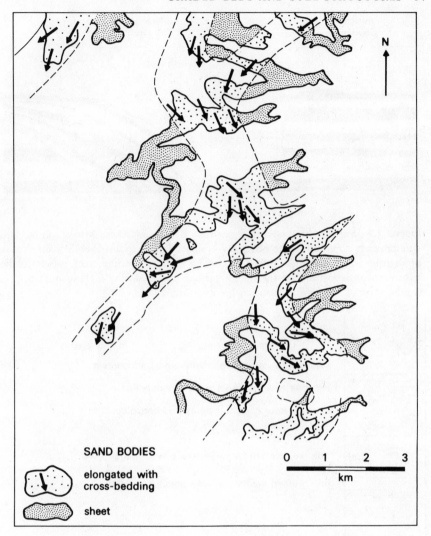

SAND BODIES

elongated with
cross-bedding

sheet

0   1   2   3
km

Figure 4.6 Channel cross-bedding. The direction of flow along, and location of, fluvial channels in the Triassic Schilfsandstein of West Germany have been inferred from the internal cross-bedding and external form of the sandstone bodies.

material is being deposited. The coarser fraction of the load is carried along at the base of turbidity currents as a traction carpet, and this undoubtedly enhances the scouring capacity of the current.

The scour marks formed in the underlying cohesive muds take on many

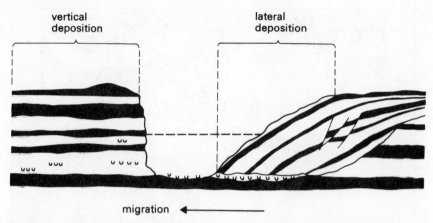

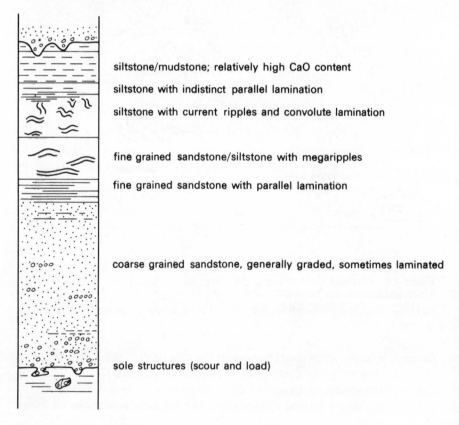

Figure 4.7 Vertical and lateral deposition in salt marshes. Microfaulting and slumping are common in the obliquely oriented lateral (point-bar) deposits. Scour of slightly firmer marsh deposits is occurring as the migrating creek moves to the left. Vertical deposition on the marsh surface occurs during high spring tides. A similar style of deposition occurs in the intertidal zone.

siltstone/mudstone; relatively high CaO content

siltstone with indistinct parallel lamination

siltstone with current ripples and convolute lamination

fine grained sandstone/siltstone with megaripples

fine grained sandstone with parallel lamination

coarse grained sandstone, generally graded, sometimes laminated

sole structures (scour and load)

Figure 4.8 (*bottom left*) Ideal sandstone turbidite unit. The thickness of these units ranges from a few centimetres to several metres. Contacts between the internal divisions may be gradational or erosional. The vertical sequence of divisions is often incomplete or partly repeated.

forms. Some are bulbous, triangular or corkscrew shaped with steep walls at the up-current end and are collectively called **flute marks** (Fig. 4.9). They grow where small irregularities have already been initiated on the mud floor, these promoting strong turbulence, eddying and increased erosion during the period of turbid flow. Other scour marks are probably eroded by pebbles, shells or wood fragments suspended within the flow (**tool marks**).

Figure 4.9 Flute and tool marks. Deltoid or triangular flutes on the underside of a Devonian sandstone bed in North America. Linear tool marks at the bottom right. The inferred direction of turbidity current flow is from top right to bottom left. (From Potter and Pettijohn (1963). Reproduced by kind permission of Springer-Verlag.)

It is probable that current directions can be deduced with reasonable accuracy from flute marks present within ancient successions. The technique has been successfully demonstrated for the Martinsburg Formation

(Ordovician) of the Central Appalachians by measuring the trend of marks at the base of graywacke beds. It is concluded that turbidity currents frequently flowed down the slopes leading away from a south-eastern land mass, then swung through a wide arc so as to travel longitudinally along elongated basins running parallel to the land mass. It is now generally recognised from such studies that longitudinal movement of sediment along the axes of deep troughs is a commonplace event during their rapid infilling. Previously, undue emphasis used to be placed on lateral input of material. A considerable proportion of the 6000 metres of Cretaceous–Oligocene flysch graywackes of the Polish Carpathians was transported longitudinally.

The rippled and laminated, finer-grained upper layers of graded units represent later phases of deposition from a decaying current. They do not usually show a marked systematic change in grain size. Fluctuating velocity and capacity of residual currents and eddies probably accounts for the variation in minor structures.

DEFORMATIONAL STRUCTURES

Post-depositional, non-tectonic deformation of sandstones and siltstones is commonplace irrespective of the environment in which the beds were formed. Many of the beds affected by deformation give every indication of a pene-contemporararaneous origin for the structures prior to the deposition of the immediately overlying bed. Others have been affected by downwards bulging of a sand bed into an underlying more plastic mud bed. Whatever the mode of origin, and this is usually obscure, the deformation appears to have occurred while the rocks were in a plastic, water-saturated state.

Distortion caused by slumping is widespread on various scales in flysch-type geosynclinal sediments and is characteristic of the flanks of geosynclinal troughs. Slumping can produce highly contorted slump sheets, balled-up masses and pebbly clay deposits which can be traced for some tens of kilometres away from the presumed source. The Eocene pebble beds of Ancon in south-west Ecuador are almost certainly of this origin, likewise the so-called tilloids of the Precambrian geosyncline of West Congo.

Some of the best exposed and most spectacular slumped beds occur in the Lower Miocene Waitemata Group of northern New Zealand. Deformation and contortion occur on a grand scale and are attributable to pene-contemporaneous slumping of poorly consolidated volcanic grits and turbidites in response to tilting of the sea floor. The contortion of the beds due to slumping closely resembles that due to tectonism so caution in inter-pretation is needed. This interpretative problem arises in most geosynclinical sequences. For example, in the Upper Ordovician flysch graywackes of Appalachia it is now believed that some 15–30 per cent of the contorted beds are of slump rather than tectonic origin as originally surmised.

Convolute lamination sometimes resembles slump bedding in internal structure (Fig. 4.10). However, the convolutions are confined to a layer within an individual bed and the layer is rarely more than thirty centimetres thick. In the upper parts of turbidite beds a layer can sometimes be traced with little or no change in thickness for several hundreds of metres. The complexity of the convolutions increases from the bottom to the top of the layer but even in the highly folded upper parts micro-faulting is absent. The structure is possibly produced by the surface drag of eddying currents of waning strength flowing over loosely consolidated sand and silty clay laminae. An origin by load deformation, simultaneous with deposition and perhaps localised by deposition on an irregular surface, is also feasible.

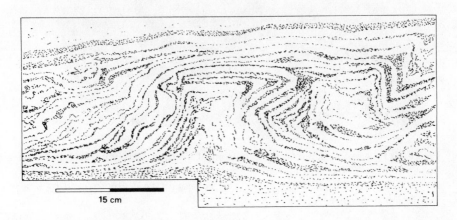

Figure 4.10 Convolute lamination in Cenozoic sandstones of the Polish Carpathians. The degree of convolution increases upwards, and the top of the convoluted layer is invariably truncated by the overlying bed.

Convolute lamination is not confined to turbidite successions and is known to occur, with less frequency, in fluvio-deltaic and littoral sediments.

Load structures generally, are disoriented bulbous projections extending downwards from the base of a sandstone bed into an underlying softer rock, normally a shale (Fig. 4.11). Although the majority of the projections have a sandstone infill equally good ones are known from the base of limestone beds. Occasionally, the structures on surfaces show a preferred orientation but close inspection invariably shows that this anomalous feature is inherited from a pre-existing linear series of flute marks. The soft mud is squeezed between the soft sand, sometimes into sharply pointed, oblique and narrow wedges referred to as **flame structures.**

Load structures are common in sediments of all successions. Some of the most peculiar deformations found in sandstone–siltstone–shale sequences are pillow- or ball-shaped masses of sandstone which are totally enclosed

Figure 4.11   Load structures on the bottom of a Lower Carboniferous sandstone bed in Illinois. These irregular structures can vary in size from a few centimetres across to a metre or so. (From Potter and Pettijohn (1963). Reproduced by kind permission of Springer-Verlag.)

in shale. They are called **pseudo-nodules** (Fig. 4.12). Internally the pillows and balls sometimes have a well developed concentric laminar structure. They are found in Upper Carboniferous beds of South Wales and northern England where they can be seen to vary in maximum diameter from a few centimetres to a metre or so. They form when a layer of sand overlying mud is subject to vibration (earth tremors) soon after deposition. The vibration causes the plastic mud to liquify and the sand layer disintegrates so that certain parts of it sink into the mud. In doing so the sunken sand parts develop saucer, boat, ball and pillow shapes depending on how much subsidence occurs.

BIOGENIC STRUCTURES

The post-depositional disturbance of sediment by living organisms is called **bioturbation.** The disturbance may be caused by organisms moving over the surface or by their penetration into the top few centimetres of the sediment, and is usually contemporary with deposition. The structures formed vary in scale and type from the surface trails (**epichnia**) of large terrestrial vertebrates to burrows (**endichnia**) of marine organisms, such as

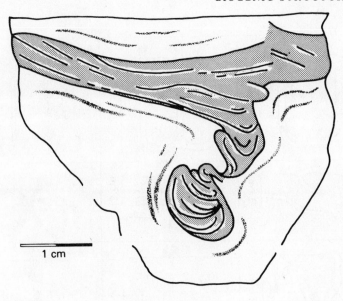

Figure 4.12 Pseudo-nodule genesis. The convoluted clayey silt nodule is just at the point where further sinking would have caused it to detach completely from its parent layer. Pseudo-nodules may become isolated from their parent layer by several centimetres or more.

Chondrites, which are a few millimetres in size. The 'hardgrounds', or overconsolidated layers, within the European Chalk are an example of churned soft sediment hardened by early diagenetic processes. The layers are about thirty centimetres thick and are full of burrows which were constructed, possibly by crustaceans, shortly before consolidation took place.

From the sedimentological viewpoint the importance of organisms which grow *in situ* or crawl, forage, burrow, rest and dwell on and in sediment is twofold. First, they physically and chemically affect the original deposit, and second they can give some slight indication of bottom conditions at the site of accumulation.

It is not always appreciated how intensive is the physical modification of soft sediment by organisms. It has been suggested that some 80–90 per cent of the comminuted shell sand fringing the Bermuda islands has passed through the gut of sea urchins and holothurians. Fecal pellets are abundant on carbonate and siliciclastic intertidal flats the world over. On some modern European inter tidal flats the degree of disturbance may be such as to destroy completely primary lami:ation and linear orientation of particles within the sediment (Fig. 4.13). *Corophium volutata*, a small but energetic arthropod often present in large numbers, is particularly adept at this. A

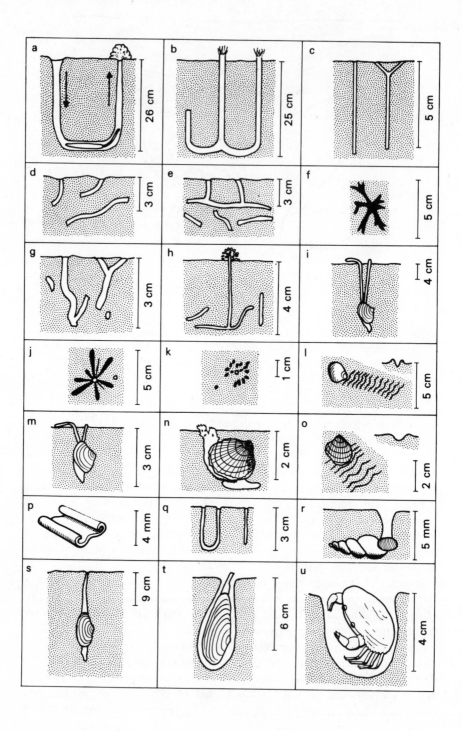

Figures 4.13 Organic structures on modern temperate zone intertidal flats. Examples from the silty and sandy muds of The Wash, east England.

(a) *Arenicola marina*, burrow; (b) *Lanice conchilega*, burrow; (c) *Pygospio* sp, burrow; (d) *Nepthys* sp, burrow; (e) *Scoloplos armiger*, burrow; (f) *Nereis diversicolors*, surface trails; (g) *Nereis diversicolors*, burrow; (h) *Heteromastus* sp, burrow; (i) *Scrobicularia plana*, burrow; (j) *Scrobicularia plana*, surface trails; (k) *Macoma balthica*, surface trails; (l) *Littorina* sp, surface trails; (m) *Macoma balthica*, burrow; (n) *Cerastoderma* (*Cardium*) *edule*, burrow; (o) *C. edule*, surface trail; (p) *Mytilus edulis*, faecal pellet; (q) *Corophium* sp, burrow; (r) *Hydrobia ulvae*, burrow; (s) *Mya arenaria*, burrow; (t) *Pholas candida*, burrow; (u) crab, burrow. After Evans (1965).

considerable amount of churning may be caused by the activities of the lugworm *Arenicola marina* when present in large numbers. Clearly the degree of biological disturbance is dependent on the density of colonisation (or persistence of ecological suitability) of a site. If the density is low, possibly because of high sedimentation rates, then the amount of disturbance is correspondingly less.

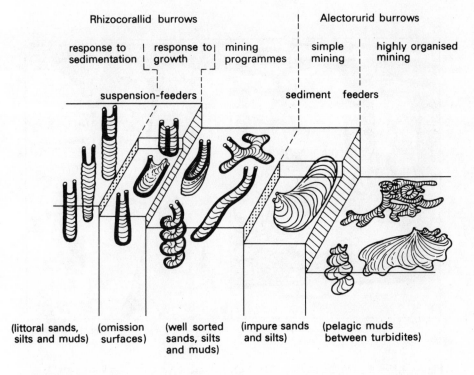

Figure 4.14 Bathymetric zonation of fossil burrows. Illustrates the general rule that suspension feeders prevail in shallow, highly agitated waters and elaborate sediment feeders in deeper, quieter waters.

The textural qualities of a sediment may be considerably modified by burrowing, with marked changes in porosity and permeability. On submarine slopes changes induced by organisms and affecting pore pressures and shear strength may be such as to make the sediment more susceptible to slumping and sliding. Conversely, burrowing activities may promote cementation and effectively stabilise the sediment.

Although the organism causing a specific type of bioturbation may be preserved in place, it is more usual to find no trace at all. In general, this creates little problem in modern deposits as there is always the possibility either that the sedimentary effects and their causative organism are known or that further biological research will identify one with the other. This contrasts with ancient bioturbation structures where the chances of successfully identifying the causes, in the absence of organic fragments, are slight. One difficulty with fossil examples is the knowledge, gained from present environments, that the morphology of certain structures can be duplicated by a variety of unrelated organisms. There is also the possibility of successive occupation of certain burrows and tubes by a number of unrelated organisms,

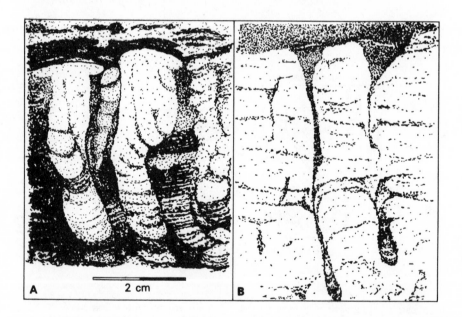

Figure 4.15 A, *Diplocraterion*. The burrows are U-shaped and internally laminated. The internal structure is caused by the upwards movement of organism in response to sedimentation.

B, *Monocraterion*. A vertical burrow pipe with a funnel-shaped upper opening.

each modifying the structure. However, the detailed study of the biological affinities and origins of the many types of ancient and modern trace fossils (**ichnofossils**) is beyond the scope of this book.

An interesting aspect of some burrows is the way in which they are distributed among particular lithological associations (Fig. 4.14).

In general, the greatest variety and highest density are to be found in deposits of marine shelf origin. The burrows are relatively simple and sometimes U-shaped, as with *Diplocraterion* (Fig. 4.15). Structurally complicated burrows, less densely distributed overall, are usually found in pelagic mud layers interbedded with turbidites, implying deeper water conditions. Then there are simple burrows, such as *Monocraterion*, which characterise terrestrial, fresh and brackish water environments (Fig. 4.15).

A careful scrutiny and assessment of the evidence provided by bioturbation structures can be useful in deducing general environments and even depths of water in which ancient sediment accumulated. But it need hardly be stressed that caution in interpretation is needed, as there is no guarantee that the structures are a completely reliable guide.

5

Rudaceous deposits

The coarser clastic sediments form a somewhat heterogeneous group. They present little of the mineralogical and mechanical uniformity imparted to many of the finer-grained clastic rocks by long-continued action of transportation, selective chemical weathering, and mechanical sorting.

In general, the pebbles or analogous components of rudaceous deposits consist of rock fragments, removed from the parent mass by mechanical

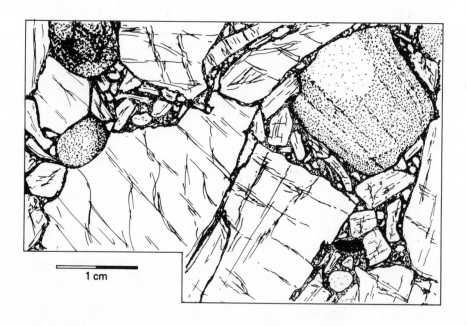

Figure 5.1 Sedimentary breccia, Old Red Sandstone, Scotland. Composed of poorly sorted, angular detritus. The fragments are predominantly vein quartz and haematised sandstone set in a fine-grained clastic matrix infiltrated by haematite.

agencies; occasionally differential chemical weathering leaves residual masses of resistant material which go to form rudaceous deposits.

CLASSIFICATION

Unlithified sedimentary rudaceous rocks, excluding the products of volcanism, are called gravels irrespective of the size and shape of their conspicuously large fragments, which are known as **pebbles, phenoclasts** or **clasts.** Lithified rocks with a dominance of rounded pebbles are termed conglomerates; when the fragments are predominantly angular they are called sedimentary breccias (Fig. 5.1). Variable degrees of roundness result in hybrid rock names such as breccio-conglomerates.

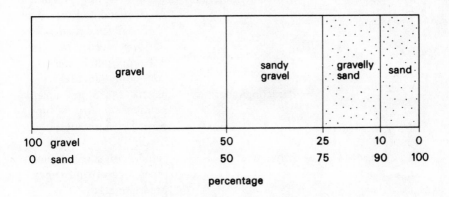

Figure 5.2 Simple classification of gravel–sand mixtures.

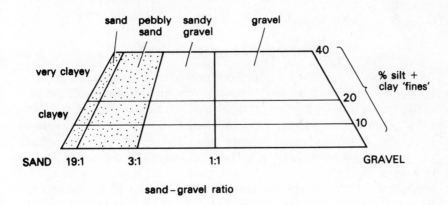

Figure 5.3 Textural classification of gravel-bearing sediments.

In theory, gravels and their lithified equivalents are distinguished from other fragmental rocks on the basis of the overall grain size being greater than two millimetres. In common practice, the names are used if the proportion of pebbles exceeds 25 per cent (Fig. 5.2).

Frequently the matrix in rudaceous rocks is either sand and silt, as in fluvial deposits, or silty clay, as in glacial and slump-slide deposits. Secondary infiltration of matrix into pores and cavities is usual. Precise description of the rock then becomes difficult, though there have been many attempts to establish suitable schemes of classification. One such, based on texture, is shown in Figure 5.3. On the other hand, if acute refinement is unnecessary, much simpler descriptive schemes can be used. Lithified rocks, for example, are capable of division as follows:

I Texture	a. Orthoconglomeratic:	framework of gravel and sand bound together by chemical cement; matrix < 15 per cent; pebbles not supported by matrix; commonly bimodal
	b. Paraconglomeratic:	matrix > 15 per cent; pebbles matrix supported; commonly polymodal
II Composition	a. Polymictic:	pebbles of several rock types, one of which may predominate
	b. Oligomictic:	pebbles of few rock types
III Source	a. Extraformational:	pebbles originate from extrabasinal sources
	b. Intraformational:	pebbles originate from within the basin

COMPOSITION OF PEBBLES

In rudaceous deposits which have been newly derived from older rocks by mechanical disintegration there is often very little alteration in the proportion of the various lithological types involved, but if the débris undergoes water transportation or chemical weathering for any appreciable length of time, the fragments of the softer and less stable rocks are reduced in size and begin to be eliminated. Thus the newly formed débris at the foot of a chalk cliff consists principally of angular pieces of chalk, with a small proportion of flint; when this material is picked up by the waves, and washed to and fro in the beach, the chalk is slowly ground away, leaving an orthoconglomeratic flint gravel. A distinctly different kind of flint gravel is produced when similar débris has undergone chemical weathering, with solution of the calcium carbonate and concentration of the relatively insoluble flint. This deposit shows no sorting or mechanical abrasion.

In common with other sedimentary rocks, conglomerates are liable to be broken down during erosion, as a result of which the pebbles of less stable rocks, such as limestones, calcareous sandstones, and basic igneous rocks in part share the fate of the cement, and become rotted during the process of weathering, so that they readily disintegrate when the material is retransported by water. Highly siliceous pebbles, however, are not eliminated in this way since they are almost indestructible by ordinary chemical weathering, and at the same time are extremely hard and tough. For this reason, once pebbles of these materials have been formed, they tend to be handed on from one conglomerate to another throughout geological time, with very little change, except possibly a slight reduction in size, or a somewhat more rounded outline. Conglomerates which have derived their pebbles at second or third hand from older rudaceous deposits without access of fresh material, tend, therefore, to contain a high proportion of such resistant materials. The principal rocks which are found to behave in this way are vein-quartz, quartzite, flint, chert, jasper, rhyolite, and quartz-aggregates derived from acid igneous rocks and gneisses.

It is seldom that a gravel is composed entirely of fragments of few kinds of rock (**oligomictic**); much more frequently it is **polymictic**, or composed of many kinds. The nearest approach to an oligomictic composition in this country is found in some of the flint gravels of the south and south-east of England, where a considerable search is sometimes necessary before a pebble of any other kind can be found.

Gravels of glacial origin, and gravels containing reworked glacial material, generally show the greatest variety of pebbles, since they often include material transported for very long distances and derived from varied sources; for example the large number of Scottish and Scandinavian pebbles in the glacial gravels of East Anglia and Yorkshire.

SHAPE AND ROUNDNESS OF PEBBLES

The shapes assumed by pebbles naturally vary according to their manner of origin. Many gravels and conglomerates contain second-hand pebbles derived from older rudaceous rocks. It is only by considering pebbles which have been newly derived from non-rudaceous rocks, and have been shaped entirely within one environment, that we can obtain any true idea of the shapes characteristic of abrasion under given conditions.

Technically it is necessary to distinguish between the shape (or form) and roundness of fragments. Shapes may be described as tabular (oblate), bladed (roller) or spherical (equant), or may be described in terms of degree of sphericity. Roundness concerns the sharpness of the corners and edges and is considered to be independent of shape (Fig. 5.4). There are five classes recognisable:

angular	little or no evidence of wear; sharp corners and edges
subangular	worn with corners and edges beginning to be rounded off
subrounded	considerable wear; corners and edges rounded to smooth curves
rounded	all corners smoothed off to broad curves
well rounded	no flat areas; entire surface consists of broad curves

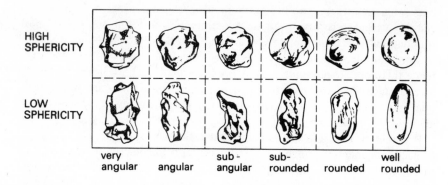

Figure 5.4 Roundness and sphericity chart for pebbles and grains in fragmental deposits.

The original shape of a rock fragment has a very strong influence upon its shape during a long part of its history, and controls the form of the pebbles even when rounding of the edges is far advanced. Thus the fissility of a laminated sandstone, or of a slate or schist, is important principally in its influence upon the shape of the original fragments, rather than in providing planes of easy fracture during abrasion. Sharp edges are worn down comparatively quickly, but flat surfaces are worn much more slowly, so that discoidal pebbles tend to retain this form even after long-continued abrasion.

The surface of a water-worn pebble is usually smooth and free from pitting or striation.

Pebbles which have been shaped during long-continued glacial transportation tend to show faceted surfaces, separated by slightly rounded rather than sharply angular edges. The flattened surfaces are frequently scratched, and the edges not uncommonly show signs of bruising. Glacial striations, however, do not develop well on all rocks. They are specially conspicuous on boulders of compact, fine-grained rocks, such as limestone, which are moderately hard, yet still soft enough to be readily scratched. In fluvio-glacial gravels the pebbles are generally much like those of ordinary rivers, but some faceted or striated blocks are usually present, the degree of rounding depending upon the amount of water transport which the material has undergone. Glacially shaped pebbles show surfaces which, apart from the characteristic bruises and striations, are smooth, and may even show a slight polish. They never show pitting or etching.

Wind action produces two entirely different pebble shapes. Small rock fragments which are not too heavy to be brought into motion by the wind, become rounded, much as they would by water transportation, and often show a remarkable approach to a spherical form. Pebbles which are too large to be transported tend to become faceted by the constant abrasion of wind-drifted sand. In this way are produced the well-known **dreikanter,** with two wind-faceted surfaces meeting in a sharp edge, the third face (base) of the pebble often being formed by part of the original surface of the rock fragment. The edges are sharp, and the abraded faces are not usually flat, but somewhat curved. Wind action produces smooth surfaces on homogeneous rocks, but if the material is irregular in hardness, the surface is pitted and etched. In all except the softest rocks the surface acquires a high polish.

The forms assumed by derived pebbles occasionally present complicated problems, but in the simpler cases are immediately suggestive of the previous history of the fragments. Boulders washed out of glacial deposits into rivers or into the sea are subjected to water abrasion, and tend to become rounded; since flat surfaces are comparatively stable in the earlier stages of water abrasion, however, the facets of ice-shaped pebbles tend to be preserved until the pebble has been considerably reduced in size.

Water-worn pebbles and boulders which are picked up by moving land ice, retain their spheroidal shape in the early stages of glacial transport, but tend to acquire faceted surfaces and striations.

SEDENTARY RUDACEOUS DEPOSITS

Where consolidated rocks are exposed at the surface of the earth to rigorous climatic conditions, mechanical weathering leads to the accumulation of loose blocks of broken rock, which are sometimes of large size. If the disintegration of the parent rock is carried out principally by frost action, the resulting débris usually consists of irregularly angular fragments, as, for instance, in the great masses of shattered rock which form the summits of many of the higher mountains in the British Isles. Where exfoliation, resulting either from temperature fluctuations or from hydration, is the dominant disintegrating process, the resulting deposits consist of more or less rounded blocks, in extreme cases showing a superficial resemblance to large water-worn boulders. This type of deposit is especially characteristic of the disintegration of well-jointed igneous rocks in arid or semi-arid countries.

The granite kopjes of Mashonaland, and the dolerite kopjes of the Karroo in South Africa, for instance, are covered with tumultuously heaped masses of huge boulders and partially rounded blocks of exfoliated rock.

Coarse-grained residual deposits are also formed by weathering from heterogeneous rocks, some parts of which are more resistant to weathering or to erosion than the rest. In Great Britain, the boulder clay frequently behaves in this way, the soft clay being readily removed from the harder glacial erratics, which then accumulate as an irregular mass of boulders.

Many of the older conglomerates possess a somewhat soft matrix or cement, so that the component pebbles are easily set free. This occurs, for instance, in the Old Red Sandstone conglomerates of Scotland, which are often very coarse in texture. Even where the cement consists of a resistant material such as silica, prolonged weathering often results in the release of the original pebbles, as, for example, at the outcrops of the auriferous conglomerates of the Witwatersrand. Concretions of harder material in soft sedimentary rocks are often left behind as a boulder deposit.

TRANSPORTED RUDACEOUS DEPOSITS

Transported deposits of coarse material are very abundant; their origin may be attributed to gravity alone, or to the action of waves, of running water, or of ice.

Scree and solifluction deposits

The screes of mountain regions consist of angular blocks of rock of all sizes and shapes that have been loosened by weathering and have fallen or slipped by their own weight. In cold regions the scree material owes its origin to the expansion of water on freezing in the joints of rocks. This process is responsible for the formation of screes in the arctic zones and in the mountains and hills of temperate latitudes, and, of course, in the highest parts of high mountains in all parts of the world. In Britain the screes of Wast Water, in the western part of the Lake District, are very well known. They rise to a height of about 450 metres above the lake in one continuous slope at an angle of about 30°. Well known, too, are the light-coloured screes on the slopes of the Dolomites in the Tirol, which from a distance look like patches of snow.

In screes which are still in course of active formation, the surface layer, especially at the upper part, is almost at the maximum angle of rest for the material involved. A slight disturbance will cause general downhill movement of the scree in the immediate neighbourhood. The deeper parts of well-established screes, however, are usually subject to chemical weathering, which allows the material to settle, with a certain amount of interlocking or cementation of the blocks, so that a considerable degree of stability is achieved. Many of the scree slopes in Great Britain date from glacial or early post-glacial times, when frost action was more vigorous.

Hillside waste which would remain at rest if the force of gravity alone were operative, may continue to move by the processes known as **creep** and **solifluction.** If a rock fragment on a slope is moved by its own expansion or that of the pore water it may be exposed to the pull of gravity. It will then creep slightly downhill. Such movements, of course, are normally infinitesimal, but since they are always directed downhill, they result in a gradual but general flow of the surface layers. Similar, but more powerful, movements (solifluction) take place as a result of alternate freezing and thawing in arctic and subarctic regions.

In certain parts of the south of England there are extensive deposits of a peculiar character, which are known as **head.** They consist of local talus or scree-like material, mixed with an earthy matrix, and sometimes showing a structure resembling bedding. These deposits are believed to owe their peculiar characters to movement by solifluction during the Pleistocene glacial period when the ground was permanently frozen to a considerable depth, as in the tundras of Siberia and North America.

In regions where great extremes of temperature occur in the course of a day, the chief agent in scree formation is the sudden expansion and contraction of the rock, the unequal stresses thus set up causing fractures. Screes are therefore well developed on steep hills and mountain ranges under desert conditions, and excellent examples can be seen in the deserts of Egypt, the Sinai Peninsula, and in many other hot and dry regions of the world. Under these conditions scree formation is much aided by a peculiar type of spheroidal weathering, for which the term **desquamation** is employed. A·notable feature of desert screes is, very commonly, the perfect freshness of the material, since chemical weathering is in abeyance, except under special circumstances.

Torrential deposits and river gravels

In the steep upper courses of rivers draining a maturely dissected country, coarse torrential gravels are transported and deposited. As the gradient of the stream becomes gentler, there is a corresponding decrease in the diameter of the particles which are transported under normal conditions of flow, and the gravels deposited are of finer texture. In the lower reaches, where the river is flowing across a well-developed flood plain, the sediments deposited on the valley-flat are usually silts and muds; gravel, if present at all, being confined to the stream bed.

The distribution of gravel deposits along the course of a river is, however, strongly modified by local conditions peculiar to each drainage basin, such as its topographical peculiarities and the seasonal or other fluctuation in discharge. Rivers which discharge from mountainous country on to flat plains usually deposit the bulk of their coarser débris in a more or less well-defined **piedmont belt,** consisting of the more or less confluent alluvial fans laid down by each stream where its velocity is checked at the foot of its steep mountain course. These accumulations come under the general heading of **torrential deposits,** and they are found for the most part in upland regions; sometimes, however, when a river has a rapid fall throughout its whole length, they may even extend down to sea level. In Britain modern and ancient examples are well developed in Wales, in the Lake District, and in Scotland.

Torrential gravels are also abundant in the valleys and on the flanks of the Alps. They are of enormous thickness, and have a very wide extension over the low ground north of the Alps and Carpathians. From a very early period to the present time their formation has been more or less continuous. Of these the fresh-water Molasse of Upper Tertiary age consists of sandstones and conglomerates formed by stream action; the Nagelfluh is a conglomerate, partly re-sorted by water; while the Pleistocene Deckenschotter includes glacial deposits and the gravels of river terraces.

Certain Precambrian Torridonian conglomerates of northern Scotland have many of the characteristics of torrential gravels, spread out on the floors of semi-arid intermontane basins. The pebbles consist of tough siliceous rocks such as vein-quartz (50 per cent), quartzites showing contact alteration, black and yellow cherts, jasper, rhyolites and spherulitic felsites similar to those of Uriconian (Precambrian) age in Shropshire. These materials are all indestructible by chemical weathering, whereas the sandy matrix, in which they are enclosed, contains an abundance of the feldspar microcline, which would be liable to decomposition by weathering in a humid climate.

The Triassic conglomerates of the eastern United States are interpreted as having accumulated in the form of piedmont gravels, brought down by torrential streams from the Appalachian Mountains. The pebbles are more or less rounded, unweathered, and, in spite of an enormous range in grain size, show little or no sorting. The material of the pebbles varies considerably along the outcrop, but always consists of easily recognisable older local rocks. In different areas the principal of these constituents are limestone, quartz, granite, gneiss, schist, and basalt, principally derived from Cambrian and Precambrian formations.

The introduction of gravel into the middle and lower reaches of relatively fast moving rivers can create gravel-rich flood plains. This happened during Pleistocene times in the Thames basin of southern England. Most of the gravel is composed of flint pebbles derived from Tertiary and Chalk sources, but some pebbles are of more exotic, far-travelled origin. During the Pleistocene oscillations in sea level, solifluxion processes and fluvial reworking variably remobilised the gravels, so that they now rest on a series of terraces and slopes elevated above the present flood plain.

Marine gravels and conglomerates

Turning now to marine deposition uncomplicated by ice, we find the conditions fairly simple. Most of the larger blocks must be obtained more or less locally; only in exceptional cases can large boulders be transported far. These cases could involve the transport of boulders off-shore by driftwood or kelp. Usually they are the result of the denudation of the coast, and their form and degree of rounding must obviously depend on many causes, the most important of which are the state of the fragment when separated from the parent rock and the amount of abrasion it has since undergone. Here, as elsewhere, boulders may be obtained ready made from an older conglomerate. One of the most notable features of all beach deposits is the amount of sorting by wave action undergone by the transportable fragments, so that on many beaches there is a regular gradation from boulders at the foot of the cliffs, through pebbles, to sand at and below low-water mark. At the foot of steep cliffs the accumulations are to a large extent of the nature of screes, but the material soon becomes rounded if within reach of wave action.

Gravels of purely marine origin are abundant all along the south coast of England, and form, in particular, two well-known deposits – the Chesil Beach in Dorset and Dungeness in Kent. The great majority of the pebbles in these deposits are of flint, and they are for the most part extremely well rounded. Since the material travels in the main from west to east (with some local exceptions), a few pebbles derived from the hard rocks of Devon and Cornwall are present. The materials of the beach gravels of eastern England, of Wales, of Lancashire and Cumberland, and of Scotland, are largely derived from glacial deposits, and their constituents are consequently of a very varied nature. For this reason they differ strikingly from those of southern England.

The older marine conglomerates are normally found as comparatively thin deposits, often associated with an erosional break or unconformity in the sequence. The majority of marine conglomerates overlie a more or less marked unconformity, and are for this reason spoken of as **basal** or **transgressive conglomerates** with respect to the succeeding strata. Other cases are known where coarse deposits have accumulated during a retreat of the sea; these are known as **regressive conglomerates.**

Marine basal conglomerates are abundantly represented amongst the sedimentary rocks of Great Britain. Perhaps the most persistent example is the basal Cambrian conglomerate, which marks the end of a long break in the stratigraphical succession. A remarkable feature is that, in many localities, the matrix contains an abundance of spherical quartz grains, mixed with angular feldspathic sand and pebbles. Now, a well-sorted sand, consisting almost entirely of spherical quartz grains, indicates that the material has had a long sedimentary history, whereas a feldspathic sand suggests recent derivation from crystalline igneous or metamorphic rocks. It seems probable, therefore, that in the matrix of such conglomerates as those underlying the Lower Cambrian sandstones of north-west Scotland or of Shropshire, we have traces of an ancient, possibly aeolian, sand, which weathered and drifted for long ages on the Precambrian land surface, until it finally became mixed with the pebbles and feldspathic débris which were freshly broken up when the Cambrian sea invaded this ancient land area.

Excellent examples of conglomerates are afforded by the coarser beds of the Precambrian Ingletonian Series of northern England. The pebbles include quartzite, granulite, schists, gneisses, phyllite and slate, together with a range of igneous rocks. It has been suggested that the beds were laid down under shallow water conditions near to a land mass, but the presence of graded bedding and sole structures in the interbedded turbidite sandstones indicates deeper waters, with possible slumping and sliding of gravel down slopes. A similar manner of deposition may also account for the Eocene clay pebble breccio-conglomerates of Ecuador. The lenticular beds are more than 200 m thick locally and consist of poorly-sorted pebbles among which clay varieties are dominant. The clay pebbles are subangular, have polished surfaces and, occasionally, have diameters of a metre or more. It is significant that the beds immediately overlying these conglomerates are of graywacke facies.

In the West Congo, Precambrian conglomerates which were originally thought to be boulder clays are now reinterpreted as sediments deposited from extensive mud slides and flows on the margins of a geosyncline. The beds have been named **tilloids** because of their resemblance to tills or boulder clays.

Boulder clays and glacial gravels

The most characteristic of all products of glaciation is boulder clay or till, which in its most typical form is a stiff clay with a varying proportion of rock boulders (Figs 5.5 and 5.6). The latter are commonly angular and have often been faceted or striated by attrition on rock surfaces, or by mutual pressure. Boulder clay sometimes contains intercalated beds of sand and gravel, and thus grades into fluvio-glacial deposits. To find true modern boulder clay we must turn to Greenland, Spitzbergen, or the Antarctic, where it seems to be a characteristic product of glaciation by continental ice sheets. Pleistocene boulder clay is largely developed in Britain north of a line joining London and Bristol, and it is also very abundant in north Germany, Holland, and the United States. As British examples the boulder clays of the Yorkshire coast may be quoted; these consist of an exceedingly stiff and tenacious clay, in which are embedded numerous angular, sub-angular, and rounded boulders whose derivation from Scandinavia, Scotland, and the Lake District is easily recognisable. These boulders are frequently of one or two tonnes weight.

Perhaps the most abundant of all glacial deposits, especially in mountain regions, are those to which the general term **moraine** is applied. Moraines consist of irregular accumulations of fragmental material, a marked feature being the indiscriminate admixture of blocks of all sizes; this is due to the fact that there can have been no sorting action as in water deposition. The blocks are commonly angular and often striated; in fact, moraine material resembles the coarser constituents of boulder clay, without the argillaceous matrix.

Very closely related to moraines are **fluvio-glacial** deposits, which consist of moraine material sorted and redeposited by running water. Here there has been usually a certain amount of grading according to size, and sometimes a rough stratification may be observed, although this is more characteristic of the finer material (Fig. 5.5). Fluvio-glacial deposits are with difficulty distinguishable from ordinary torrential boulder deposits; but the inclusion of far-travelled blocks sometimes affords a useful index to their origin. It is perhaps worth noticing here that the so-called **anchor ice** of northern rivers often transports large boulders, and mixes them with other deposits either lower down the river or even in the sea. A similar process is the carrying of large quantities of material over the sea by floating icebergs derived from the land. This certainly accounts for many of the large boulders found in marine

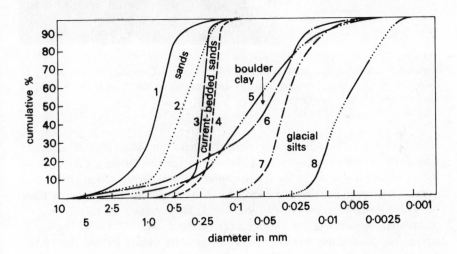

Figure 5.5 Mechanical composition of glacial and fluvioglacial deposits. All samples are from North America. The boulder clays (5,6) show characteristic poor sorting. The remainder, which are much better sorted, show the effects of water transport to varying degrees. The silts (7,8) have a well marked maximum limit to grain diameter, but a less distinct lower limit; this is a feature of mixed grain size material which has settled from suspension.

sediments; and in the Arctic and Antarctic regions it gives rise to a peculiar type of oceanic deposit. The total amount of material carried south by icebergs during Pleistocene times and dropped on the Dogger and New-foundland banks, for example, must have been enormous.

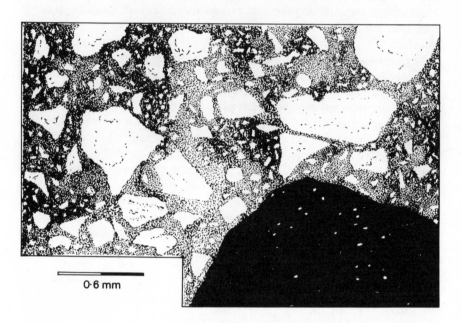

0·6 mm

Figure 5.6   Boulder clay matrix, Pleistocene, Cheshire, England. The angularity of many of the quartz grains and the unsorted nature of the deposit are typical of modern and ancient boulder clays.

Instances are known in the far north where ice laden with boulders is driven on to the shore, or forced through a narrow channel, and both of these causes may result in the formation of boulder terraces along the coast.

Pre-Pleistocene boulder clays are known by the name of **tillite**, and have been recognised in many parts of the world. They contain irregularly shaped or angular boulders of fresh, unweathered rock, with all sizes of pebbles mixed indiscriminately together in a fine-grained matrix of angular chips. The Upper Carboniferous tillites of the southern hemisphere are perhaps the best-known examples (e.g. the Dwyka Tillite of South Africa). The Talchir Boulder Bed of India is of similar age.

The Dwyka Tillite forms the base of the Karroo System north of latitude 33°, and rests upon a striated pavement of older rocks. The boulders, which are often faceted and scratched, include an immense variety of rocks derived from pre-Karroo formations, such as basic lavas, granites, gneisses, grits, jaspers, slates, sandstones, and quartzites. These are set in a tough, bluish-grey matrix of angular sand mixed with argillaceous material. This deposit reaches a thickness of over 300 metres, and for the most part has all the characters of an indurated boulder clay. There are, however, at some localities beds of conglomerate with a calcareous cement, which appear to represent fluvio-glacial gravels, and bedded shales are also found, occasionally showing a varve-like banding characteristic of clays laid down in glacial lakes.

Tillites of late Precambrian age have also been reported from many parts of the world, such as Canada (Huronian Tillites) and South Africa (Government Reef Series). The Dalradian Port Askaig, Kilcherenan and Loch na Cille conglomerates of Ireland and the western Highlands of Scotland may be tillites.

Intraformational conglomerates and breccias

The conglomerates which have already been discussed in this chapter all contain pebbles or similar bodies derived from older consolidated rocks, and there is a considerable difference in age between such pebbles and the matrix which encloses them. There are, however, other rudaceous rocks in which this age distinction is slight or negligible; deposits of this kind are termed **intraformational conglomerates** or **breccias**, and in certain environments are formed in the normal course of continuous sedimentation, without the intervention of diastrophism and subsequent erosion. Thus their stratigraphical significance is entirely different from that of most other rudaceous rocks.

Streams flowing over an alluvial plain tend to shift their channels laterally. In doing this, the streams erode the recently deposited sediments of which their banks consist, and large masses of coherent mud and silt, released by undercutting, fall into the channel and are rolled along its bed. Transported fragments of this kind come to rest on the inner, slack-water, side of the meander, and become buried in sand and mud. In times of heavy flood, vigorous erosion of the banks takes place, and the débris may be broken up into small, semi-plastic pellets which are distributed by the flood water over the surface of the plain. Deposits of this kind, known as **clay-gall conglomerates** (with sandy matrix) and **shale conglomerates** (with argillaceous matrix), are of frequent occurrence in the Coal Measures and Wealden deposits of Britain, and are also conspicuous in a somewhat similar facies of the Middle Jurassic sediments on the Yorkshire coast.

Mud-flake conglomerates are also formed as a result of the exposure of large areas of newly deposited mud to the atmosphere with consequent desiccation of the surface layers. They are found in a wide range of environments. In semi-arid and desert countries, for example, large pools or lakes, often no more than a few cm in depth, are formed by seasonal rains. When the water disappears again and the freshly deposited sediment is exposed to the atmosphere, the surface layers shrink on drying, and a system of polygonal desiccation cracks appear. These polygonal masses of mud are sometimes picked up by the wind, and accumulate elsewhere to form an intraformational breccia, or they may remain, becoming hardened by further desiccation and baking in the sun, until they are buried under later sediments or removed by flowing water during a later flood.

6

Arenaceous deposits

Sediments of arenaceous or sandy texture are characterised by grain sizes which fall on the Wentworth scale between $\frac{1}{16}$ and 2 mm. For the rocks described in this chapter the term siliciclastic is also appropriately descriptive.

Quartz is the commonest sand-forming mineral, and may be regarded as the principal constituent of most arenaceous rocks. Many sands consist almost entirely of quartz, but frequently appreciable proportions of feldspar and white mica are present. Other components are subordinate, and are only discoverable by a microscopic examination of the residue after the removal of the quartz, feldspar, and mica. They consist of such resistant minerals as were able to survive the weathering processes that effected the destruction of the rocks in which they originally occurred. These are known as **accessory** or **heavy minerals**, some of the most common of them being ilmenite, rutile, garnet, tourmaline, zircon, staurolite, kyanite.

When quartz grains are freshly liberated from a previously lithified rock, whether igneous or sedimentary in origin, they retain many of the characters which they possessed in the parent rock. Quartz, however, is an extremely durable mineral, and individual grains are passed on from one sandy deposit to another through long periods of geological time. During this process, the external shape of the grains undergoes repeated modification, and the original characters are progressively obscured and obliterated (Figs 6.1 and 6.2). At the same time, the less stable minerals, originally associated with the quartz grains, are constantly being eliminated, so that sands of ancient derivation which have not received additions of more recent material gradually approach the condition of being pure silica deposits. In studying any particular arenaceous rock it is important to bear in mind the possibility of the grains having been derived from an older sandstone, and to distinguish between the age of the grains and the age of the deposit.

Quartz grains derived from granites usually present an irregularly angular but roughly equidimensional form. When the mineral is liberated by chemical decay of the surrounding feldspars, the grains have a frosted and corroded surface, due to the solvent action of alkalis liberated from the decomposing feldspars. A similar effect can be seen in grains loosened from a calcite (alkaline) matrix. Quartz liberated by mechanical destruction of undecayed granite is sharply angular, with clean, glassy fractures. A study of the inclusions in detrital quartz grains may often give an indication of the source of the material; the inclusions in the quartz of granites most commonly consist of small crystals of other minerals, such as prisms of tourmaline and needles of rutile. The quartz of acid lavas are more apt to contain glassy inclusions.

Mica-schists and thinly foliated gneisses, when newly broken up, are liable to produce quartz grains which are flattened in the plane of foliation, and which may consist of a group of crystals rather than a single large individual. The grains are referred to as polycrystalline.

Coarsely banded gneisses, on the other hand, give quartz grains very much like those derived from granites. Gneisses derived from intensely metamorphosed sediments not uncommonly supply quartz grains characterised by inclusions of sillimanite in slender needles with a felted or plume-like arrangement.

Granules of quartz derived from fine-textured siliceous rocks, such as flint or chert, are readily recognised between crossed nicols, when the minute constituent crystals extinguish individually to give a highly characteristic minutely speckled appearance.

Great care has to be taken to avoid provenance misinterpretation because quartz grains may be recycled, extensively reworked in the basin of deposition or transported over long distances. During all these processes the polycrystalline grains appear to be selectively disintegrated by mechanical and chemical agencies.

Undulatory or strain-shadow extinction, a common feature of quartz grains, is of doubtful value in provenance studies as many igneous and metamorphic rocks contain quartz of that type.

The rough, irregularly angular grains derived from weathered crystalline rocks, provide the raw material for most arenaceous deposits (Fig. 6.2). On being picked up by rivers, this material is to some extent sorted according to grain size, but retains its subangular character, even after long transportation. Beach sand, unless derived from older sandstones, is also subangular, but the constant to-and-to movement under the action of waves very slowly rounds off the sharper edges of the grains. However, the grains rarely become well rounded in these situations (Figs 6.3 and 6.4).

Some of the best rounded of modern sands are found in deserts, such as those of north Africa (Fig. 6.5), and are believed to owe their shape and smoothness to abrasion during long-continued transportation by wind. The grains affected are usually larger than $0 \cdot 1$ mm in diameter. Below that size a marked degree of angularity may be retained. Similar well-rounded sands are found in the Permo-Trias beds of north-west Europe.

Figure 6.1 Sand isolated from the Eilean Dubh Dolomite, Ordovician, north-west Scotland. Remarkably well rounded quartz grains of this kind have a wide distribution in the early Palaeozoic sediments of North America and north-western Europe. The rounding may have been accomplished on the old Precambrian land surface before the sand was transported into the Palaeozoic seas. The grains are up to one millimetre in diameter.

Figure 6.2 River sand, Recent, Scotland. The grains have been derived at first hand from crystalline schists, and show the sharply angular outlines characteristic of grains in their first cycle of sedimentation. The grains are up to half a millimetre in size.

Figure 6.3 Beach sand, Recent, southwest England. Some of the grains are angular, but most of them show a moderate degree of rounding. This sample is typical of the beach sands of southern England, which are mostly derived from pre-existing sandstones, and are in at least their second cycle of sedimentation. The grains in this sample are derived mostly from Devonian sandstones. The largest is half a millimetre in diameter.

Figure 6.4 Wind-blown sand from dunes behind the beach at Gruinard Bay, north-west Scotland. The grains are derived from various sources, including gneisses, Precambrian and Mesozoic sandstones and Pleistocene glacial deposits; consequently they show varying degrees of rounding slightly modified by wind-action. The main characters, however, are still those of a beach sand. The grains are up to half a millimetre in diameter.

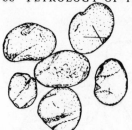

Figure 6.5 Wind-blown sand, Nubian Desert, Sudan. The grains are in an advanced state of rounding and all traces of angular edges have been removed, except where minute chips have been recently flaked off the rounded grains. The mechanical composition of this sand is shown in Figure 2.6. The smallest grain is 0·5 millimetre in size.

A common surface feature of the grains is frosting, a micro-pitting caused either by wind abrasion or the precipitation of a vitreous film of silica from capillary waters. It has been suggested that chemical corrosion by alkaline capillary solutions can create the same effect in desert situations. On the other hand, in environments where water is more freely available and alkaline in nature, secondary chemical frosting can occur during cementation. This affects grains of all shapes and degrees of roundness. Marine calcareous sandstones show it and so do calcrete- and caliche-type sandy soils.

Electron microscope studies on unlithified sand deposits show that particular styles of transport and deposition impress certain types of pitting on grain surfaces. The distinction between and evaluation of new and inherited textures is difficult. In ancient sandstones the environmental diagnosis value of these studies is diminished by cementation processes, which can destroy depositional surface textures.

CLASSIFICATION

The arenaceous deposits include ordinary unconsolidated sands, such as those found on sea-beaches and in rivers, and their lithified equivalents, the sandstones. The range of dimensions of grains to which the term sand is commonly applied will be seen by reference to Table 2.1 on page 16. Sand grains usually consist of individual mineral fragments, though composite particles, derived from very fine-grained parent rocks, are not uncommon. In most unconsolidated sands, the pore space between the grains is occupied by air or water, but in environments where sorting processes are not active, the sand grains may be embedded in a matrix of silty or argillaceous material. The grains may be welded together under pressure, without addition of cement from outside, and the pore space becomes reduced or disappears in the process. Open sands may become lithified by subsequent deposition of some substance between the grains, this secondary binding material being termed a **cement**. In the case of sands with a matrix already present between the grains, induration of this material may serve to lithify the deposit without the introduction of a cement.

Many schemes of classification for the arenaceous rocks have been drawn up, some based upon theoretical principles, some emphasising the mineral composition, and others devised empirically for convenience in field and laboratory description.

Current tendencies are to subdivide sandstones on the basis of their mineral content, or mineral and textural attributes. None of the proposed schemes is used universally, so that care is needed when interpreting the nomenclature used by various authors. Some of the attributes of sandstones may be easily recognisable in the field but, almost invariably, require confirmation by ancillary microscope work. Cements play no role in sophisticated modern classifications, though matrix minerals, which equally well may have been introduced or be diagenetic in origin, are frequently taken into account.

In this text the classification adopted is that of Dott modified by Pettijohn, Potter and Siever in 1972 (Fig. 6.6). Two main groups of sandstone, namely arenites and wackes, are recognised and separated on the basis of their matrix (< 30 microns) content. Arenites, which have less than 15 per cent matrix, and wackes, which have more, are then subdivided on the grounds of various mineral and rock particle attributes.

Unfortunately, one of the consequences of adhering to any particular scheme is the inevitable disappearance of some well-established names. One such is orthoquartzite which, in the proposed classification, falls into the quartz arenite category. Protoquartzite and subgraywacke, terms with a

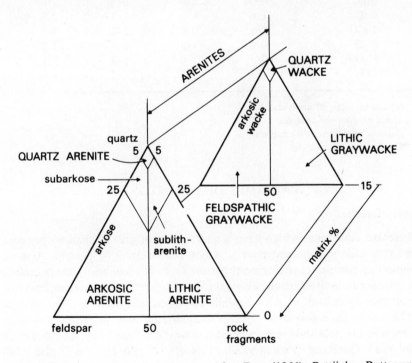

Figure 6.6 Classification of sandstones. After Dott (1964); Pettijohn, Potter and Siever (1973).

confusing history of definition are now replaced by sublitharenite and lithic arenite respectively.

As might be anticipated the mineral and textural parameters used in classification are reflected by the chemistry of sandstones. Quartz arenites have a very high silica content and low alumina content, arkoses have a comparatively higher potash and alumina content, and so on (Table 6.1). Whether looked at from a chemical or mineral point of view it is necessary to appreciate that the rock as eventually constituted reflects the composition of the source rocks, the nature and maturity of the weathering processes, the quantity and quality of diagenetic changes and the presence or absence of biochemical or other contaminants.

Table 6.1 Mean chemical composition of main sandstone classes.

	1	2	3	4
SiO_2	96·0	66·1	75·5	63·7
Al_2O_3	0·8	8·1	10·4	14·2
Fe_2O_3	0·3	3·8	1·7	1·5
FeO	0·2	1·4	0·8	4·5
MgO	0·04	2·4	0·5	2·6
CaO	1·6	6·2	1·7	2·9
Na_2O	0·1	0·9	2·4	3·7
K_2O	0·1	1·3	4·1	1·4
H_2O+	0·3	3·6	0·9	3·1
CO_2	1·1	5·0	1·2	1·0

1 Quartz arenites, 35 analyses.
2 Lithic arenites, 20 analyses.
3 Arkosic sandstones, 38 analyses.
4 Graywackes, 81 analyses.

VARIETIES OF SANDSTONE

Quartz arenites

These are sandstones which have a quartz content greater than 95 per cent and very little fine-grained matrix. A few pebbles may be present. A small amount of the quartz is commonly in the form of secondary quartz cement in optical continuity with the original clastic cores. In some ancient beds low grade metamorphism may have mildly enhanced this effect.

The rocks have also been called orthoquartzites and quartzites, the latter being a long-established term perpetuated stratigraphically by such names as Holyhead Quartzite (Precambrian of North Wales) (Fig. 6.7), Hartshill Quartzite (Cambrian of central England) and Eriboll Quartzite (Cambrian of north-west Scotland).

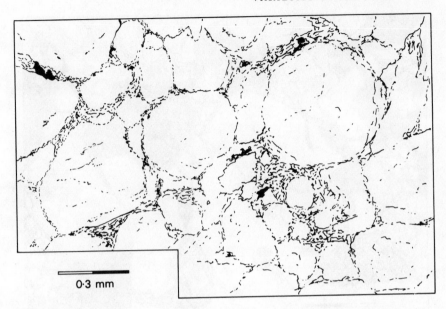

Figure 6.7 Holyhead Quartzite, Precambrian, North Wales. A moderately well sorted quartz arenite with the original subrounded grains modified by authigenic outgrowth. The margin of the original clastic cores is faintly picked out under plane polarised light by an intermittent film of clay minerals. The small amount of matrix is formed predominantly of illite flakes, but there are minor patches of secondary limonite. Polarised light.

The high detrital quartz percentage almost certainly reflects a recycled origin from a deeply weathered quartz-rich terrain, plus an additional amount of winnowing-out and attrition of more labile minerals during contemporary transport and deposition. Sorting is very variable. It is hardly surprising that these sediments accumulate in quantity in littoral, beach and dune zones.

Under conditions of prolonged marine transgression, quartz arenite sands may be smeared in extensive sheet-like and lensoid bodies above the plane of marine erosion. In Britain good fossil examples are found in the Lower Palaeozoic strata of the Welsh Borderland and include the Wrekin Quartzite (Cambrian) and Stiperstones Quartzite (Ordovician) (Fig. 6.8). Fresh bedding surfaces of these hard, but brittle, rocks have a light gray to white lustre. Ripple marks and bioturbation structures are common at some levels.

When warm water currents prevail and soluble calcareous skeletal matter proliferates, the transgressive sand bodies may become rich in introduced calcareous cement as lithification proceeds. The proportion of cement might

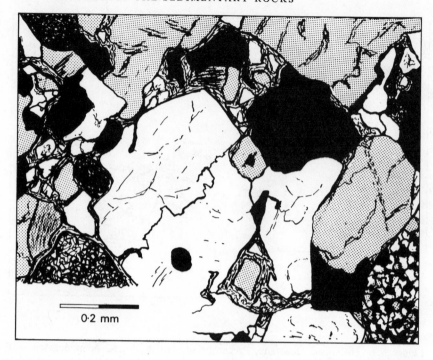

Figure 6.8 Stiperstones Quartzite, Ordovician, Welsh Borderland. A quartz arenite consisting of an ill sorted aggregate of rounded to subangular quartz grains and lithic fragments of sandstone and siltstone. The angularity of the quartz grains is mainly due to authigenic outgrowth. The matrix consists of illite, chlorite and carbonate. Crossed polars.

reach 40–60 per cent so that the textural characteristics become atypical of quartz arenites. The name calcarenaceous sand (or sandstone) has been suggested for this special variety.

Some modern quartz arenites are recorded in relatively deep basinal waters and may either have been swept from shallow shelves or may be in-place, current-winnowed relict deposits of earlier periods of lower sea level. There is also the possibility of deposition from melting ice.

Terrestrial varieties of quartz arenite include those of wind-blown origin, such as the Penrith Sandstone (Permian) of northern England with its well-rounded grains (Fig. 6.9), and those produced by in-place weathering, such as ganisters.

Ganister, a name originating in northern England for a sandstone of refractory quality within coal-bearing successions, consists of angular and subangular quartz grains cemented by secondary quartz (Fig. 6.10). Early diagenetic kaolinite occupies interstitial cavities. The feldspar and clastic

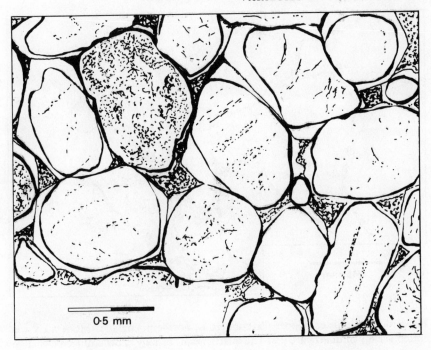

0·5 mm

Figure 6.9 Penrith Sandstone, Permian, north-west England. A medium-grained, relatively well sorted quartz arenite. The authigenic enlargement of the quartz grains is emphasised by a thin, but dense, pellicle of haematite and clay minerals around the original rounded cores. The outgrowths of clear quartz are in optical continuity with dusty cores. The cement is predominantly iron oxides. Polarised light.

mica content is low compared with associated sandstones. Traces of roots may be present and plant débris may be locally abundant. The originally deposited sand appears to have been subject to leaching and this is likely to have occurred during a phase of subtropical weathering and soil production.

Some types of silcrete are pebbly quartz arenites produced during surface weathering, probably under semi-arid climatic conditions in contrast to ganister formation. Cementation is either by authigenic quartz outgrowth or by introduced opaline and chalcedonic silica. Post-Cretaceous red, buff and brown coloured sedimentary silcretes are known in Britain, America, South Africa and Australia.

Lithic arenites and sublitharenites

These sands and sandstones constitute a major group and are common in

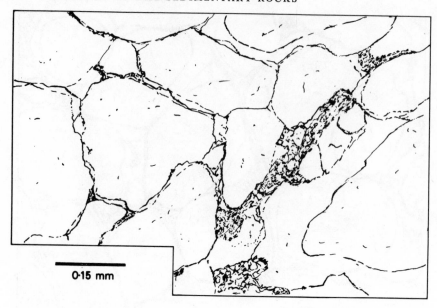

0·15 mm

Figure 6.10 Ganister, Upper Carboniferous, northern England. Moderately well sorted variety of quartz arenite. The quartz grains usually show authigenic enlargement and intergrowth; a secondary angularity is developed. The matrix is predominantly kaolinite and illite. Polarised light.

geological successions. They include the protoquartzites and subgraywackes of some authors.

The quartz content in lithic arenites (or litharenites) varies up to 75 per cent, rock fragments are dominant over feldspar grains, and the matrix, which is mainly clay of mixed detrital and authigenic origin, forms less than 15 per cent of the rock. Sublitharenites are a subgroup straddling the boundary between lithic arenites and quartz arenites.

Introduced cements commonly include secondary quartz, calcite, siderite and iron oxides and hydroxides. In most instances the quartz, calcite and siderite have been precipitated early during diagenesis and siderite, in particular, may even be a primary precipitate. The carbonates often show corrosive relationships to the clastic grains they are partly enclosing, this being detected by micro-embayments at the periphery of the quartzes and cleavage infiltration in feldspars and micas. Limonite is widespread as cement and is usually attributed to secondary superficial changes.

The detailed make-up of the clastic grains is very variable depending on the constitution of the provenance and the environment of accumulation. Modern alluvial and estuarine sands in south-east England contain quartz, chert, flint, vein quartz, quartzite, feldspar, glauconite, barytes and clay

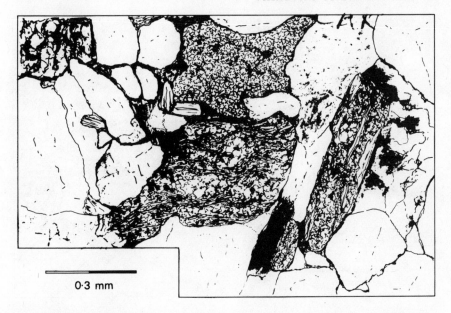

0·3 mm

Figure 6.11 Farewell Rock, Upper Carboniferous, South Wales. A rather poorly sorted lithic arenite consisting of angular to subrounded grains of quartz, vein quartz, silty shale and slate set in a small amount of matrix, consisting of clay minerals and secondary limonite. Crossed polars.

grains derived from adjacent Pleistocene and Tertiary outcrops. This material is admixed with present day mud, silt and shell fragments and set in a mud matrix. The labile constituents such as the shells are progressively lost during lithification.

Lithic arenites of Lower Old Red Sandstone age in the Welsh Borderland have Lower Palaeozoic volcanic rocks, argillaceous limestones and siltstones as their main lithic constituents. These rocks, which are part of a thick terrestrial succession, exhibit fluvial characteristics. Terrestrial successions the world over comprise to a large degree lithic arenites. The Tertiary terrestrial sandstones of the northern Mediterranean average 25 per cent quartz, 10 per cent feldspar, 25 per cent limestone grains and variable amounts of chert and volcanic fragments.

Many of the fluvial and deltaic sandstones of Upper Carboniferous age in the northern hemisphere are lithic arenites or sublitharenites (Fig. 6.11). They are interbedded with, and grade laterally into, quartz arenites. Clastic mica, dominantly muscovite, is frequently concentrated along thinly spaced bedding planes and produces a flaggy structure. The cementing agents are commonly kaolinite, illite, siderite and limonite.

The fluvial Pennant sandstones (Upper Carboniferous) of South Wales

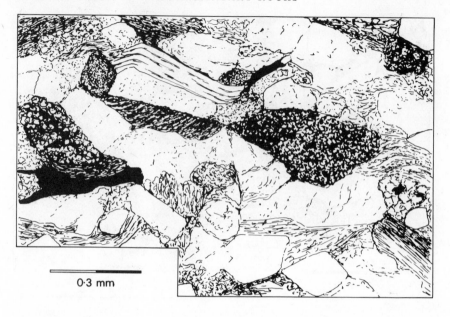

0·3 mm

Figure 6.12 Pennant sandstone, Upper Carboniferous, South Wales. A poorly sorted lithic arenite composed of subangular grains of quartz, metaquartzite, muscovite, silty shale and slate set in a matrix of illite, kaolinite and carbonaceous matter. The muscovite grains and matrix are often squeezed and distorted due to compaction. The quartz grains sometimes show marginal intergrowth with the matrix. Polarised light.

are lithic arenites in part. They contain about 10 per cent matrix, less than 10 per cent feldspar and 25–55 per cent rock fragments. The fragments are dominantly intrabasinal Devonian and older Carboniferous sediments, the rest being extrabasinal pre-Upper Palaeozoic igneous and low-grade metamorphic rocks (Fig. 6.12).

In addition to fresh-water and sea-margin sequences, lithic arenites are reported as occurring frequently in post-Jurassic marine basin successions.

An inherent property of all lithic sandstones, irrespective of environment of accumulation, is their high potential for provenance determination. Rock fragments are more rewarding than individual crystals when inferring the constitution of source areas. Schist, slate and tuff fragments, for instance, are much more informative than undulatory quartz grains. Moreover, if the rock fragments are weak and susceptible to rapid physical disintegration, it is reasonably certain that they have not passed through several cycles of erosion and deposition.

Arkosic arenites and related rocks

This group includes arkosic arenites proper and subarkoses.

Sands freshly derived from undecomposed crystalline rocks, such as gneisses or granite, contain notable quantities of feldspar, even in humid climates. If the feldspathic sediment is deposited almost at once, and sealed off from circulating ground water, the feldspar is preserved, even though weathering, erosion, and deposition have all taken place in a humid climate.

Work on the petrology of the Californian Coast Ranges has shown that the arenaceous rocks of all ages from the Jurassic onwards to the present day are prevalently feldspathic rather than quartzose. Orthoclase and acid plagioclase frequently make up half or more of the sediment by weight, and sometimes are present in considerably greater abundance. These rocks, having a feldspar content greatly in excess of 25 per cent and matrix less than 15 per cent are arkoses in the strict sense. In fact, the clay matrix in the Tertiary beds rarely exceeds 4 per cent, though carbonate cement, which is discounted for classification purposes, ranges up to 40 per cent.

The quartz of the Cretaceous arkoses contains fluid inclusions suggestive of derivation from granite, and the accessory minerals, green hornblende, magnetite, zircon, sphene, apatite, and others, provide further evidence of this origin. In parts of the Tertiary, additional minerals such as glaucophane and brown hornblende indicate that much of the sediment was supplied by rivers draining an area of Franciscan metamorphic rocks, a view which is confirmed by a study of the accompanying conglomerates. We thus have an imposing sequence of Mesozoic and Tertiary arkoses, derived directly from mildly chemically weathered crystalline rocks. The sands were rapidly transported and those with a low matrix and high carbonate content were probably deposited in a marine environment.

In the arenaceous rocks of western Europe, feldspar is not found to be so persistently abundant as in the region discussed above, but important deposits of arkose appear at certain horizons. In parts of the Torridon Sandstone (late Precambrian) of Scotland, feldspar is almost as abundant as quartz. In spite of the great age of the sediment, the dominant feldspars, microcline and microcline-microperthite, are remarkably fresh and angular (Fig. 6.13). The matrix of these sandstones is usually a mixture of clastic and authigenic clay minerals into which red iron oxides have infiltrated. The overall impression is of very limited transport and minimal sorting by moving water, even though cross-lamination and cross-bedding are widespread. Terrestrial deposition is implied.

Fresh microcline is well represented in the Millstone Grit (Namurian) sandstones of northern England, a deposit laid down in the deltas of a river system draining a distant northern land area. Certain beds are genuine arkoses, but more have a 5 to 25 per cent feldspar content and are more appropriately referred to as subarkoses. In contrast to the Torridon beds,

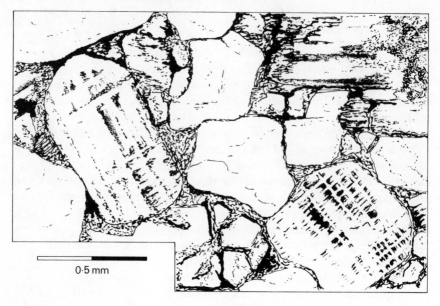

0·5 mm

Figure 6.13 Arkose, Torridonian, north-west Scotland. A poorly sorted rock consisting of subangular to rounded quartz, vein quartz and metaquartzite grains, and a relatively high proportion of fresh microcline, microperthite and acid plagioclases. The matrix is commonly rich in illite and introduced iron oxides. The quartz grains sometimes show marginal intergrowth with the clay minerals. Crossed polars.

the arkosic and subarkosic sandstones carry little clay matrix and have a relatively greater proportion of secondarily introduced chemical cement (Fig. 6.14).

The Precambrian sparagmites of Scandinavia consist mainly of coarse arkoses and subarkoses, containing high proportions of microcline, and a little plagioclase. The earlier sparagmites are usually gray in colour, but the later beds are often bright red. There is no evidence for regarding these rocks as having been formed under desert conditions and there are reasons for considering them as products of a cold, and possibly subarctic, environment. On the other hand, some of the upper sparagmites in Sweden have been claimed as reflecting a change to a warm, humid climate.

It may be pointed out that marine arkosic rocks are found as basal deposits, formed by the reworking of granitic debris. This type of arkose is developed near the base of the Cambrian succession in England wherever the underlying rocks are of suitable composition. The lowest beds of the Malvern Quartzite (otherwise a quartz arenite) where they rest on Precambrian gneisses provide a typical British example of a marine arkose.

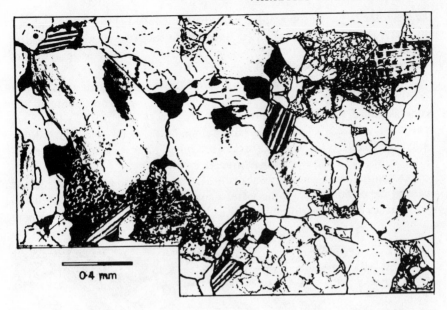

0·4 mm

Figure 6.14 Subarkose, Upper Carboniferous, north England. Slightly better sorted than the Torridon arkose (Figure 6.13) and has more secondary cement (ferroan carbonate). Some of the quartz grains show evidence of authigenic enlargement followed by corrosion by carbonate. Secondary interstitial kaolinite is present, often abutting fresh microcline and plagioclase feldspars. Crossed polars.

Graywacke sandstones

These sandstones, which are various shades of gray in colour, are more difficult to define satisfactorily than any other group of sandstones. This is because their mineralogy tends to be complex with angular to subrounded quartz, plagioclase and rock particles set in a finer-grained matrix. The matrix is difficult to resolve under the microscope, but generally consists of a mixture of silt- and clay-sized quartz, feldspar, illite, montmorillonite, chlorite and mixed-layer clay minerals. These are intermingled occasionally with epidote, pyrite and carbonates.

The micaceous minerals are partly derived from the breakdown of unstable silicates in the source areas, and are clastic. The rest are almost certainly diagenetic, forming after the deposition and burial of the sand. Loading and subsequent tectonism, and mild metamorphism, accelerate the post-depositional processes of change of the unstable grains. The progressive consequences of such change are particularly evident in many pre-Tertiary graywackes, where grain boundaries become diffuse because of intergrowth between the larger clastic grains and matrix minerals. Feldspars are replaced by secondary zeolites and epidote. At a certain stage the clastic plagioclases begin to convert into Na-rich albite. The Na_2O content of graywackes frequently exceeds 3 per cent (Table 6.1).

The classifications that take into account the matrix indicate that it must form more than 15 per cent of the rock. When feldspars are dominant the rock is called a feldspathic graywacke (Figs 6.15 and 6.16); when rock particles, such as volcanic ash, are dominant the rock is called a lithic graywacke (Fig. 6.17). Quartz wackes are much rarer rocks and tend to be interbedded with 'cleaner-washed' quartz arenites.

Feldspathic and lithic graywackes are commonly associated with marine basinal deposition, though the water depth need not be great. Many lithic graywackes are rich in volcanic detritus and seem to have been laid down in

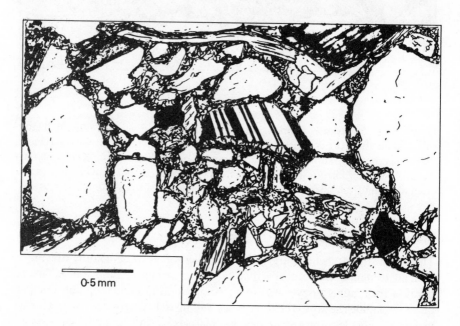

0·5 mm

Figure 6.15 Feldspathic graywacke, Cambrian, North Wales. Feldspar grains are dominant over rock fragments in this very poorly sorted sandstone. The bulk of the feldspar is fresh and angular. Many of the grains show marginal intergrowth with the silty clay matrix, in which illite is a prominent component. Crossed polars.

Figure 6.16 (*top right*) Feldspathic graywacke, Devonian, Germany. A classic example of graywacke consisting of a mixture of subangular to subrounded grains of altered and fresh feldspar, quartz and a range of rock fragments including metaquartzite, vein quartz, slate, tuff and trachyte. Marginal intergrowth of the clastic matrix of illite, chlorite and quartz with the larger grains is usual. Crossed polars.

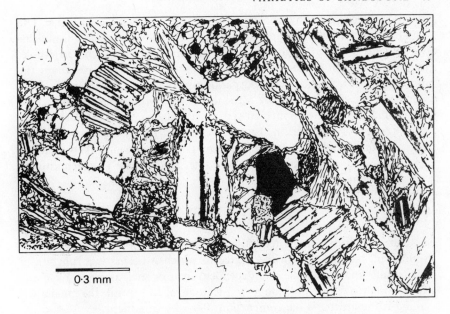

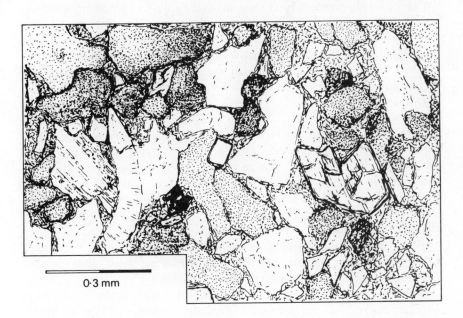

Figure 6.17 Lithic graywacke, Silurian, southern Scotland. Angular highly degraded rock fragments, partly sericitised feldspar, quartz and occasional pyroxene (right centre) set in a relatively small quantity of matrix. Small grains of epidote (centre) are present in small amounts. Polarised light.

troughs adjacent to active volcanic island arcs near plate margins. The Tertiary Waitemata Group graywackes in northern New Zealand are of this type and carry abundant andesitic and basaltic fragments. On the whole, graywackes carrying such rock fragments are more informative about the nature of the provenance than feldspathic equivalents.

Graywackes characteristically form the lower parts of repetitive, graded beds found in thick geosynclinal-type successions (Fig. 4.8). The poorly sorted texture of the sand in these beds reflects, to some extent, mass deposition from fast-flowing turbidity currents. But it seems likely that the poor sorting is emphasised by a variable amount of post-depositional recrystallisation.

DIAGENESIS

Compaction

The compaction or volume reduction of sands is predominantly caused by overburden loading, though the degree to which it occurs is a function of many small- and large-scale interdependent factors, such as grain shape and size, sorting, packing, mineralogy and tectonism. In general, the greater the loading is the greater is the compaction. Porosity is reduced.

On deposition, some well-sorted coarse sands may exhibit a very open, primary packing (or spatial arrangement) of the grains, with porosities verging on 40–50 per cent. After mild loading and readjustment of the grains to a tighter packing, the sand might then have its porosity reduced to 25–30 per cent. If smaller grains were originally introduced into the sediment, so as to occupy interstitial positions between the larger grains, then the loading might reduce the porosity to some 15 per cent or less.

In the top few metres of newly deposited, highly porous and well-sorted sands the pore pressures may build up to exceed the loading pressures until a point is reached when any type of small vibration triggers off mass movement. The sand then flows or liquifies, fluid is dispersed and the grains readjust to a tighter configuration. Sometimes, because of high rates of deposition in rapidly subsiding basins, sand may be buried so quickly that liquifaction does not occur. The sands then may remain overpressured at depth, with relatively high pore pressures, such that they never compact to the expected amount.

Under more usual conditions, compaction proceeds by individual grains being pushed closer together. Whereas an originally deposited sand may have grains which have small or tangential point contacts without interpenetration, a buried loaded sand may have a much higher proportion of closer, long, concavo–convex and microstylolitic (sutured) contacts. It has to be appreciated, however, that the original shape of grains determines to some extent the nature of the contacts, irrespective of subsequent loading adjustments.

At greater depths of burial or during phases of marked tectonism, quartz grains begin to slide over each other, fracture and split at contact points, and hence adopt a closer packing. Feldspars break even more easily, whereas micas, clay minerals and lithic fragments bend and slide acting, together with small amounts of pore fluid, as lubricants. Arkosic sands (15–25 per cent feldspar) are much more compactible than quartz arenites (5 per cent feldspar). Clay-rich and lithic sands, e.g. graywackes, may achieve up to 40 per cent compaction. By the time that plastic deformation, extensive shearing and granulation occurs, without necessarily increasing compaction, the processes pass into the realms of metamorphism.

Cementation

Sands with open pore spaces are normally lithified by the introduction of a cement, deposited from circulating solutions, or formed by the redistribution of some original constituent such as calcium carbonate or colloidal silica. The proportion of cement required to lithify a sand is comparatively small (5–10 per cent), and the quantities involved could in many cases be supplied by redistributive processes acting within the sand itself, or by the solutions expelled from adjacent argillaceous or calcareous rocks during compaction. For expelled solutions to be effective agents of cementation the water must be in large quantity, free-flowing and maintain a reasonable degree of chemical uniformity for a long period of time.

The role of water, the usual fluid expelled, needs some emphasis as there is a tendency to underplay its importance in many diagenetic studies. Whether fresh or marine, meteoric, connate or juvenile the water is an active and changing entity, which has a prominent, if not dominant, effect on the course of diagenesis. Reactions and inter-reactions are constantly taking place between the water and the materials with which it comes into contact and, as a consequence, the chemistry of the water changes, albeit slowly. There may be a desalting of the water or, conversely, a concentration of salts, or a preferential increase or decrease of some dissolved constituents. In fact, the changes may be such that the present chemistry of the interstitial waters, if determinable, may bear little relationship to their original chemistry. The pore waters in marine strata, presumably originally derived from the sea, show a marked deficiency in Mg and marked increase in Ca compared with sea water. There are differences also in the Na/K and Ca/Na ratios. Certain of these differences may be caused by ionic filtration processes which seem to be particularly effective when interbedded argillaceous sediments are compacted and a proportion of the salts are retained by the clay. Other changes probably involve adsorption, ionic exchange, pH–Eh relationships and microbial activities.

Some authorities have concluded that cementation of sands can only take place above the water table, basing the argument upon the assumption that active circulation of ground water (i.e. meteoric water) is essential for the operation of this process.

A certain style of superficial cementation takes place actually at the surface in dry countries, or in countries with a prolonged dry season. Where the water table lies near the surface, water is drawn upwards by capillarity to make good the loss by evaporation, and if any suitable material is carried in solution, the surface layers (duricrusts) become cemented. In South Africa, for example, near-surface cementation of this kind produces extremely tough surface quartzites (silcretes), limestones (calcretes or caliche) and ironstones (ferricretes) according to the nature of the solutions present. The precipitation of gypsum in salina and sabkha areas of the Middle East produces gypcretes.

Concretionary cornstones are ancient varieties of caliche formed by soil-producing processes under semi-arid conditions. Good examples occur in the Old Red Sandstone of the Welsh Borderland, Scotland and Ireland, and in the lower part of the Carboniferous successions of Scotland.

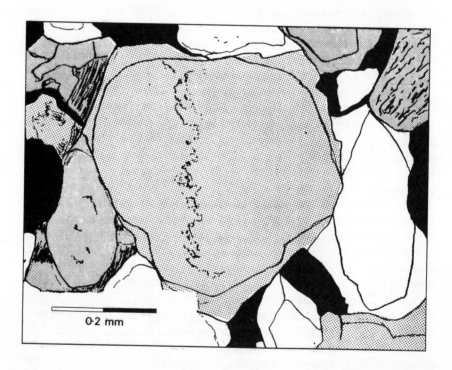

Figure 6.18 Quartz cementation, Eocene, Mississippi. A quartz mosaic formed by authigenic enlargement of clastic quartz grains. The original cores are indicated by a continuous line of 'dust' particles, commonly clay minerals. At certain points of contact the original cores show partial intergrowth, probably as a result of pressure solution effects. Crossed polars.

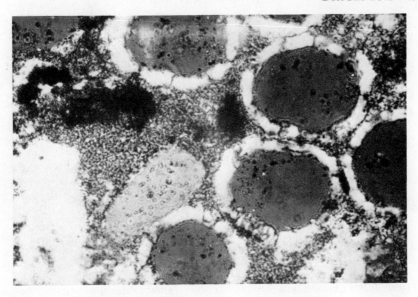

Figure 6.19 Cementation by small quartz grains. Small clear crystals of secondary quartz forming fringes around detrital grains. This is an unusual effect the reasons for which remain obscure. Crossed polars, × 50.

Diagenesis does not bring about the conversion of quartz into other minerals, but it is very common to find opal and chalcedony changing into quartz. The change often goes hand in hand with cementation, the opal and chalcedony of the matrix altering and merging progressively into authigenic quartz outgrowths marginal to the original quartz grains. The quartz cement of sandstones is usually deposited in optical continuity with the detrital grains (Figs 6.18 and 6.21) but may also form fringes of minute crystals growing radially outwards from such grains (Fig. 6.19) or the original pore space of the rock may be filled up by an irregular mosaic of minute crystals.

In the deposition of secondary quartz around original nuclei of sand grains, the crystalline continuity of the new and old quartz can be demonstrated by their simultaneous extinction when a thin section is rotated between crossed polars (Figs 6.20 and 6.21).

Sandstones showing this texture may be met with in deposits of all ages. The Bunter sandstones of northern England, which locally consist of well-rounded quartz grains formed by wind action (the so-called millet-seed sand), show a secondary quartz growth in crystalline continuity round original grains coated with hydrated ferric oxide.

Where the cementation has been continued sufficiently long, the cemented

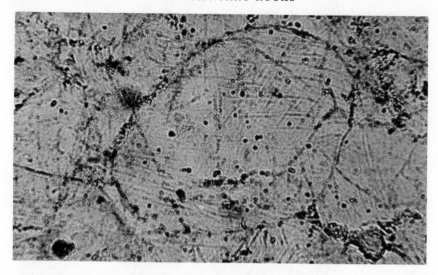

Figure 6.20 Cementation by enlargement, Lower Palaeozoic quartz arenite, north-west Scotland. The secondary quartz, which completely fills the original pore-spaces of the rock, is in crystalline continuity with the detrital quartz grains. The straight sutures between many of the enlarged grains result from the preservation of newly formed crystal faces. Polarised light, × 100.

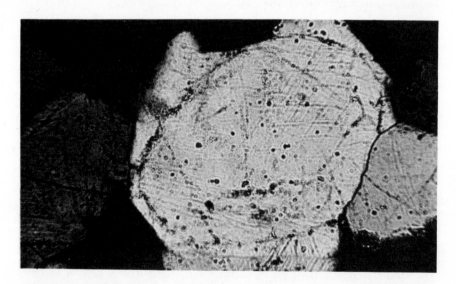

Figure 6.21 Cementation by enlargement. Same slide as Figure 6.20 but photo-graphed under crossed polars to bring out more clearly the optical continuity between the cement and the detrital grains. The rounded outlines of the original grains are marked by traces of dark impurities. Crossed polars, × 100.

grains fit so closely together that they mutually interfere and prevent the development of crystal faces.

The problem of the origin of secondary or authigenic quartz cements has been discussed at great length but there is little general agreement. Some suggest that the new quartz can be produced quite adequately by the pressure solution of the original clastic quartz. The clastic quartzes dissolve at their points of contact and the silica in solution is immediately precipitated around the grains in positions of lower pressure. In contrast, others suggest that the solution of interstitial grains of quartz (less than 0·02 millimetres in size) may create 'siliceous fluids' within the loose sand and from these fluids secondary quartz cement is precipitated. Devitrification of volcanic glass in sands almost certainly contributes additional silica for quartz cementation; likewise the leaching of silica from clay minerals distributed in the sand. An important source for marine sands is siliceous organisms such as diatoms, radiolaria and sponges. On death their chemically unstable opaline silica skeletons dissolve, enriching the pore waters in silica and this becomes immediately available for precipitation as secondary quartz.

Precipitation of calcite cement depends largely on increasing the carbonate–bicarbonate ratio in the interstitial waters. This is accomplished either by increasing temperatures or by increasing pH, both of which decrease the solubility of calcite.

Sandstones are often cemented by calcite which, if secondary, usually post-dates secondary quartz cements in the same rock. The quartz grains in these rocks sometimes develop a secondary angularity due to reaction between the carbonate-rich pore waters and the grains. In thin section calcite appears to be corroding the quartz.

The proportion of calcite to detrital quartz varies considerably, as also does the arrangement of crystals in the cement. In lightly cemented sands, such as some of the Pleistocene sands of Britain, there is merely a fringe of minute calcite crystals coating the quartz grains, cementing the rock near the original points of contact, but leaving large voids elsewhere. There is every gradation between this type and its completely cemented derivative in which the pore spaces have become completely filled by granular calcite.

In a calcareous sand, consisting of quartz grains and shell-chips, the pressure due to the weight of overlying sediment is carried at the points of contact between adjacent grains. The solubility of calcium carbonate is increased at these points, and the material of the shell-chips goes into solution, to be redeposited as crystalline calcite in the open pore spaces. Many calcareous sandstones appear to have become lithified in this way. In some cases the cement in each interspace is a single crystalline unit, the crystals in adjoining spaces being independently orientated. The most remarkable structure is found in those rocks where the cement is in conspicuously large crystals, each enclosing a considerable quantity of detrital quartz sand. On broken surfaces of the rock, the cleavage surfaces of the cement give a characteristic lustre, and each crystal unit independently gives a more or less brilliant reflection when held in an appropriate position. This phenomenon is known as **lustre mottling**. The large cementing crystals have irregular, wavy, or slightly interlocking outlines, but in a few cases the cement occurs in large euhedral crystals which can be readily isolated from the less perfectly cemented sand in which they are embedded. The best-known European occurrence is that in the Oligocene Fontainebleau Sands of the Paris Basin. The calcite here is in the form of rhombohedra, embedded in comparatively loose sand. Thin sections show that the detrital quartz gains within the cement are not in contact with each other.

Ferruginous cements (mainly haematite, turgite and limonite) are common in red bed successions (Fig. 6.22). The iron oxide is often present as an investing pellicle around each quartz grain, and where silicification has subsequently taken place the iron coating often becomes enclosed by the new crystal growth. On the other hand, the secondary quartz itself sometimes becomes coated with iron oxide, and successive periods of ferruginous and siliceous cementation may be thus distinguished. The turgite ($2 Fe_2O_3.nH_2O$),

Figure 6.22 Ferruginous sandstone, Lower Carboniferous, Scotland. The angular and subangular quartz grains are loosely packed and many appear to be 'floating' in the dense secondary haematite and limonite cement. In fact, the bulk of the grains are supporting each other by contacts of the tangential-point type. Polarised light, × 30.

which appears to be the main red colouring cementing agent in the Penrith and St Bees Sandstones (Permo-Trias of north-west England) is probably primary and formed around clastic grains in the basins of deposition, where semi-arid climatic conditions prevailed. Some of the turgite seems to have been transported in from adjacent wetter highland areas, although the bulk seems to have been precipitated from capillary waters in the basin sediments.

The term 'red beds' is normally used in describing thick, reddened terrestrial clastic successions, often of molasse-type. Some of the oldest occur in the Precambrian Torridonian succession of north-west Scotland, the Sparagmite Formation of Scandinavia and the Keweenawan beds of the Lake Superior region. All the geological systems from Cambrian to Tertiary are known to have major 'red bed' occurrences. Probably the best known in the northern hemisphere are of Devonian (Old Red Sandstone facies) and Permo-Triassic (New Red Sandstone facies) age.

The Fe_2O_3 content of red sandstone and siltstone beds rarely reaches more than 5 per cent and is frequently as low as 1 per cent. Not all the Fe_2O_3 is present as haematite, some almost certainly forming part of the chemical structure of the other grains, more especially clay minerals.

Siderite (chalybite) is probably much more often present in the cement of sandstones than is commonly realised. In some cases there can be little doubt that the mineral has replaced original calcite. An important series of sandstones in which siderite is the sole cement has been described from the Lower Lias of Sweden. The rocks consist of fine-grained, clean sand, enclosing unaltered marine fossils, and the sideritic cement is in the form of minute yellowish crystals. On weathering, the siderite is principally converted to red–brown limonite. The secondary red staining created by post-Carboniferous pre-Permian weathering of Coal Measure siderite-rich sandstones in north-eastern England is predominantly due to the alteration of siderite.

In the British Isles barytes has been described as a cement in the Triassic Keuper sandstone in the neighbourhood of Nottingham. The barytes is present as a microcrystalline aggregate either uniformly distributed or aggregated in streaks and patches. In places it amounts to as much as 50 per cent of the rock, occurring in patches which preserve their optical continuity over large areas, and thus produce a kind of 'lustre mottling'. It is probable that in most cases the original cementing material was not a sulphate, but a carbonate (witherite) deposited from water in which it was dissolved as a bicarbonate, and that the sulphate has been formed by a metasomatic interchange with soluble sulphates.

Instances are known of the formation of isolated crystals of barytes containing from 40 to 60 per cent of quartz sand. Such crystals and groups of crystals are found in the Nubian sandstones (Carboniferous–Cretaceous) of Egypt.

Gypseous sandstone, in which the cementing material is gypsum, is usually characteristic of cementation under arid and semi-arid conditions. Such deposits have a wide distribution on the modern sabkha areas of Middle East countries and have also been described from the Russian steppes and from Bolivia. In all these cases the gypsum tends to form large crystalline aggregates, sometimes with definite crystal shapes, enclosing as much as 60 per cent of sand. Some of the aggregates are called 'desert roses' because of their distinctive petaloid morphology.

ACCESSORY MINERALS

In addition to such minerals as quartz, feldspar and mica, sands contain small quantities of other minerals some of which have formed in place (authigenic) and others of which are detrital.

Glauconite

Glauconite is particularly widespread in marine sandstones of all ages and, occasionally, is present in sufficient quantities to impart a green tinge to the fresh rock. The Lower Cretaceous greensands of south-east England, which are predominantly sublitharenites, are prime examples (Fig. 6.23). At weathered outcrops the green tinge is usually obscured by brown-red staining partly caused by the breakdown of the glauconite, but not entirely as even the most deeply weathered sections still contain fresh grains of the mineral. Weathering leads to a progressive loss of potassium and the production of a montmorillonite or vermiculite structure. Iron is released so that the glauconite grains have the appearance of having degraded to limonite.

Glauconite comprises a group of green minerals all of which are potassium iron silicates. Glauconite proper is an iron-rich clay mineral with a well-ordered, high potassium, mica-type lattice with less than 10 per cent expandable layers. A disordered, non-swelling, mica-type lattice and a disordered, swelling, montmorillonite-type lattice are characteristic low potassium 'glauconites', whereas other 'glauconites' consist of an admixture of two or more clay minerals. In swelling varieties the expandable layers may be over 50 per cent.

Although most glauconites form in marine alkaline situations (pH 7–8) there is now ample evidence for deposition in mildly alkaline waters of fresh-water lakes and other continental environments.

Greensands in modern oceans originate near the edge of the continental shelf, though the mineral glauconite may be transported after formation into both deeper and shallower water. The principal deposits are situated in warm waters (15–20 °C) along the south and east coasts of Africa with an especially strong development on the Agulhas Bank; along the south and west coasts of Australia; along the west coast of Portugal; off the Californian coast, and along the edge of the continental shelf off the east coast of North America between Cape Hatteras and the Bahamas Banks. The tops of off-shore

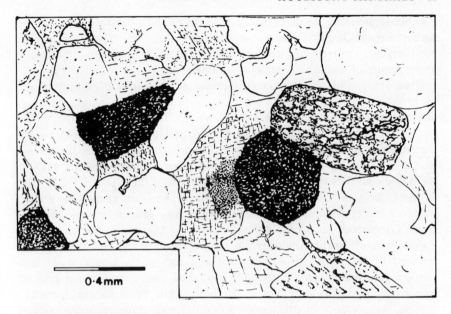

Figure 6.23 Glauconitic sandstone, Lower Cretaceous, south-east England. A calcareous sublitharenitic variety of sandstone in which the glauconite grains (dense stipple) vary from olive-green to pale greenish-yellow in colour. Some may be partially converted into secondary limonite. Several of the glauconite grains appear to have been reworked and rolled along the sea bottom soon after formation, whereas others, of a more diffuse patchy nature (centre) have not. The quartz grains show marked evidence of corrosion by percolating carbonate-rich pore waters, probably an event almost synchronous with the precipitation of the coarse sparry calcite cement. Polarised light.

Californian banks, between depths of 40 and 500 metres, are particularly rich in glauconite with percentages rising to 20. The waters in these areas are relatively quiet and oxygenated, though it is probable that the glauconitisation processes occur in mildly reducing situations (Eh 0 to -200 mV) in the top few centimetres of the sediment. Formation is facilitated by the presence of decaying organic matter.

Glauconite is sometimes found filling the chambers of foraminifera. The material of the shell does not take part in the chemical reactions of glauconite formation, but merely provides an enclosed space in which special reactions may take place.

There is much evidence to support the belief that glauconite may be formed from certain clay minerals, especially if those minerals have a degraded layer silicate lattice. Potassium and iron appear to be progressively adsorbed onto the lattice, which probably has a low lattice charge, with the production of glauconite.

Other suggestions are that glauconite forms by the breakdown of minerals such as amphiboles, pyroxenes, feldspars and biotite mica. There is evidence, however, that the reactions involved are not so much breakdown, but more in the nature of actual replacement by glauconite.

Glauconitic sands are abundantly represented at various levels throughout the geological column. They are often associated with breaks in sedimentation and marine transgressions. The Lower and Middle Cambrian of the Welsh borderland contain beds of greensand, and similar rocks of approximately the same age are found elsewhere, as in Scandinavia. Glauconitic limestones have a wide distribution in the Lower Ordovician of certain regions, such as the Baltic countries and North America. In most of these Lower Palaeozoic deposits, the glauconite occurs principally as water-worn or angular grains. Evidence of deposition in very shallow water may be seen in some of these rocks. Greensands are strongly developed in the Lower Cretaceous both of Europe and of North America, the mineral being usually in the form of rounded or botryoidal grains. In the Gault and in the Chalk of western Europe, glauconite occurs in variable quantities, usually as internal casts of sponge spicules or foraminifera. A few horizons in the Chalk are character-ised by the presence of green-coated nodules with a surface impregnation of glauconite. Greensands and green clays are also well developed in several formations in the Eocene of Great Britain, as for example in the Bracklesham Beds of the Hampshire Basin.

Both ancient and modern glauconitic deposits commonly contain phosphate pellets and pebbles which suggests very slow clastic sedimentation at the time of deposition.

Heavy minerals

Accessory constituents are usually scattered throughout the rock, but may be concentrated for study. The majority of these mineral shave specific gravities greater than that of quartz, and are commonly separated by allowing them to sink in appropriate liquids whilst quartz and feldspar are floated off. The resulting concentrate is commonly spoken of as a **heavy mineral residue**.

The hard and stable minerals of regional metamorphism, such as **garnet**, **kyanite**, and **staurolite**, are common accessory minerals, as also are some of the minerals of pneumatolysis, such as **tourmaline** and **topaz**. The accessory minerals of granite, for example **zircon**, **sphene**, and **monazite**, also have a wide distribution. The minerals of purely thermal metamorphism appear to be less stable as detrital grains than are those of large scale thermodynamic metamorphism, and such minerals as **andalusite**, **sillimanite**, and **cordierite** have a distinctly restricted distribution. Other common detrital minerals are **rutile**, **apatite**, and **ilmenite**; somewhat less common, though widely distri-buted are **corundum**, **anatase**, **brookite**, **chloritoid**, and the **spinels**, **pyroxenes**, and **amphiboles**.

When they are newly derived from crystalline rocks, the grains of these minerals tend to be comparatively unworn, and consist of cleavage forms, angular fragments, or more or less complete crystals. If the material becomes involved in a second cycle of sedimentation, the content of accessory minerals usually suffers some change, even though no new material is added. Unstable minerals tend to be eliminated, or to be replaced, by new alteration products. Such minerals as olivine, hypersthene, augite and brown hornblende are particularly susceptible to alteration under sedimentary conditions, and would not be expected to survive more than one cycle of sedimentation. At the other extreme, hard and stable minerals such as zircon, tourmaline, and rutile are almost indestructible, and are passed on from one sediment to another through many cycles of sedimentation. Such grains in time become smoothed and worn, but the process is very slow, and the well-rounded grains found in some sedimentary rocks indicate that the individuals in question are of very great geological age as detrital grains. The preponderance of zircon, tourmaline, and rutile in the heavy residue of a sand almost certainly indicates derivation from pre-existing arenaceous rocks, especially if the grains of these minerals show appreciable signs of abrasion.

Since the accessory minerals of a sand are principally survivals from the parent crystalline rocks from which the sediment was derived, and the character of the assemblage of these minerals depends upon the history of the sediment, it is often possible to learn much of this history by a careful study of the minerals present. In particular, the accessory minerals usually yield evidence of the source of provenance of the sediment, either by indicating directly some specific terrane, or by ruling out those areas which could not have supplied the assemblages found, and thus narrowing down the possibilities. By using this method of investigation it is often possible to obtain information which is of considerable service in reconstructing the geography of the areas which supplied the basin of deposition. It is necessary to understand that suitable allowances have to be made in all deductions for size-sorting and specific gravity-sorting of the heavy minerals during their transportation. Moreover, post-depositional processes may alter the constitution of heavy mineral assemblages under certain solution and weathering conditions.

Work on the New Red Sandstone (Permo-Trias) of the west of England illustrates the use of accessory minerals in tracing the source of arenaceous sediments. The heavy mineral assemblage in these beds was found to be large and varied, but two important sets of minerals could be distinguished, one derived from a region of granitic intrusions, and the other from an area of regional metamorphism. Blue tourmaline, topaz, garnet, cassiterite, fluorite, rutile, and brookite suggest derivation from the contact-metamorphic and pneumatolytic rocks and the minor intrusions associated with the granite masses of Devon and Cornwall. The minerals of regional metamorphism, such as staurolite and kyanite, on the other hand, could not be derived from the same area, and the metamorphic rocks of Brittany provide the nearest available source. Thus we may picture sands and torrential gravels being swept into this Permo-Triassic desert basin from a massif of metamorphic rocks lying to the south, and also from the region of the granite intrusions lying to the west.

The assemblages of accessory minerals present in the sands of a long persistent basin of deposition usually vary not only within each bed when traced horizontally along its outcrop, but also vertically from one formation to another. If the horizontal variations are slight when compared with vertical changes in mineral composition, the latter may be used as a basis of correlation over restricted areas. This method has been found to be of considerable service in recognising horizons in thick series of unfossiliferous sands found in some oilfields, and thus providing a means of correlation between the sections in different wells. Correlation by this means is not always possible and even where it is practicable, considerable caution should be used in interpreting the results.

The causes of local variation in the heavy mineral content of a sandy deposit are several. In the first place, sediments from different sources may be brought independently into the basin of deposition, for example by rivers, by glaciers, or by coast erosion with or without littoral drift. Even in basins supplied by a single drainage system, the distribution of accessory minerals is not, as a rule, quite uniform, since some minerals are less easily transported than others, and thus tend to remain concentrated near to the source of supply. The differences found between the mineral contents of contemporaneous sands in closely neighbouring basins are sometimes very great indeed. For example, the marine Middle Jurassic sands of the English Midlands contain garnet, staurolite, kyanite, glaucophane, chlorite, chloritoid and topaz, and have quite evidently been supplied by rivers draining a region of metamorphic rocks. Further north in Yorkshire, the Middle Jurassic sands contain an extraordinary abundance of the titanium minerals rutile, ilmenite, anatase, and locally brookite, the only other important minerals being zircon and tourmaline, often in much-rounded grains. An assemblage of this kind strongly suggests derivation from older sediments, and not from metamorphic or igneous rocks.

Vertical variations in the heavy mineral content of a series of superimposed sand formations may arise from changes in the direction of littoral drift, from the arrival of sediment supplied by a different drainage system, or from changes in the existing system, such as the exposure of deeper-seated rocks as denudation progresses.

Variations of this kind are strikingly displayed by the Upper Palaeozoic rocks of Britain. The earliest Carboniferous deltaic sandstones of eastern Scotland contain a mature assemblage of subrounded heavy minerals (zircon, topaz, rutile, tourmaline, anatase, and monazite) suggestive of derivation from pre-existing sediments. These sediments were probably Lower Palaeozoic and Precambrian in age and mantled a complex core of metamorphosed rocks which was only exposed by deep erosion in later Carboniferous times. The evidence for this latter conclusion rests with the presence of a more variable heavy mineral assemblage (garnet, magnetite, ilmenite, zircon, topaz, rutile, xenotime, monazite, tourmaline and brookite) in the later and more wide-spread deltaic sandstones. There is also a wide range of metamorphic and igneous pebbles in some of the later beds. The source of the deltaic materials throughout was a land area to the north and north-east of the Midland Valley of Scotland.

Beach concentrates

Concentrates of dark sand on the upper part of the beach are familiar features of sandy shores. These beach sands consist of heavy mineral grains of high specific gravity, selectively concentrated by wave action from the ordinary beach sands, in which they were originally present as accessory minerals. Their composition naturally varies from place to place; in the north-west of Scotland, ilmenite concentrates form sands of a dark bluish-black colour. The brown sands of the east coast of Yorkshire consist in large measure of partly oxidised grains of siderite rock, mixed with a few polished grains of oolitic limonite and a great variety of minerals derived from the boulder clay. Further south, for example on the East Anglian coast near Hunstanton, these limonite ooliths, derived from Cretaceous and possibly other ironstones, form a remarkable concentrate of dark-brown, highly polished grains.

In various parts of the world, similarly formed concentrates are developed and sometimes contain high proportions of rare minerals, which may be of commercial importance. The beach concentrates of ilmenite, rutile, and zircon in Florida have been extensively worked as a plentiful source of zirconium and titanium. Monazite sands are of economic importance since this mineral is a source of the rare earths, especially thorium. Monazite is a widely spread accessory constituent of granitic rocks, and in certain localities, especially Brazil and southern India, it has accumulated in large quantities in the sands of the sea coast, having been naturally concentrated by wave action, along with garnet, zircon, ilmenite, magnetite, and other heavy and resistant minerals.

Around the shores of the Pacific, and in the beds of some rivers draining into this ocean, large deposits of 'black sand' are extensively developed, and are of some commercial importance. The beach concentrates consist chiefly of magnetite and ilmenite, and other heavy and stable minerals, but they are chiefly notable for containing workable quantities of gold, with sometimes platinum and diamonds. They are extensively developed in Alaska, Idaho, Washington, Oregon, and California, and in New Zealand; in all these localities they have yielded much gold.

Placers

Alluvial or **placer** deposits containing precious accessory minerals are of considerable commercial importance in many parts of the world.

Alluvial sands and gravels accumulating in the valleys of rivers and in lakes, to which they have been transported by running water, at one time yielded the bulk of the world's output of gold. The gold is concentrated in the coarse gravels and among the boulders at the bottom of the placers, the most valuable accumulations being often actually on the bed-rock itself. If the latter happens to consist of steeply dipping schists or slates, the upturned edges of the latter act as natural riffles or bars to catch and retain the gold particles. Accumulation also takes place on what are known as 'false bottoms', which are beds of clay or sand cemented by iron ('pan'), alternating with the beds of gravel.

Placer gold is usually associated with a heavy black sand consisting of magnetite, ilmenite, and haematite, together with chromite, garnet, zircon, spinel, and other heavy resistant minerals; but obviously the particular association is determined by the nature of the parent rock. The character of the gold is very variable: it occurs in flat scales and flakes, in rounded particles, and as irregularly shaped grains and nuggets bearing evidence of much attrition. Crystallised gold also occurs. Placer gold varies in size from the finest dust to nuggets weighing thousands of grams; the larger nuggets having received much of their gold by accretion within the gravel since deposition.

Some of the richest gravels are formed by a secondary mechanical concentration or re-sorting of earlier auriferous gravels, which today are found as terraces tens of metres above the present river beds, as in California.

Placer deposits occur in the river systems of every part of the world; but the greatest amount of gold has been won from the Recent and Pleistocene gravels of California, Alaska, Australia, and Siberia. The older gravels are often deeply buried under a thick cover ('overburden') of clay, soil, peat, and moss, which is sometimes permanently frozen, as in the tundras of Siberia and Alaska. In California and Australia the gravels of the ancient river systems are also often concealed beneath later flows of lava, and are then known as 'deep leads'.

The platiniferous placers of the Iss and other rivers draining the eastern slopes of the Ural Mountains in Russia deserve special mention. In these deposits the platinum is associated with chromite and magnetite; and their origin has been traced to intrusive masses of peridotite. Platinum is also occasionally associated with gold in the residual and alluvial gravels of California, British Columbia, Brazil, Colombia, and Borneo.

By far the larger proportion of the tin ore of the world is obtained from alluvial and other deposits. Cassiterite (SnO_2), the only important ore-mineral, is very hard and heavy, and highly resistant to weathering; it therefore tends to concentrate naturally in all superficial deposits derived from tin-bearing granites, in which it occurs as a primary mineral, and to an even larger extent from the associated pneumatolytic and metamorphic rocks. Cassiterite may be concentrated in workable quantities in residual deposits left by differential weathering at the outcrops of tin-bearing rocks, but more commonly it has been transported some distance from the original source. As in the case of the gold placers, the concentrates tend to accumulate at the base of the gravel, and to rest upon the underlying floor or bed-rock.

The early tin output of south-west England, which goes back almost to prehistoric times, was almost entirely from alluvial deposits. Much alluvial tin has also been obtained from Nigeria and eastern Australia.

Gem-gravels and sands in which, among other precious stones, diamond is found are also of economic importance. These interesting accumulations owe their origin to a concentration of the same nature as that which produces the auriferous gravels – the difference in the material accumulated being solely dependent on the local occurrence of a parent rock containing the minerals characteristic of the deposits.

Diamantiferous gravels in which the diamonds are sufficiently numerous for profitable extraction are not common. Those of the Vaal River in South Africa are perhaps the best known. The diamonds are found in residual and alluvial gravels derived from lavas and the Carboniferous Dwyka conglomerate.

7
Argillaceous deposits

The fine-grained clastic sediments include clays, shales, mudstones, marls and the aeolian deposits which originated as dusts. Shales are finely laminated rocks in which the individual parallel laminae represent periodic phases of slow sedimentation in a low energy environment. The rock is fissile, splitting easily along the planes of lamination. Mudstones are blocky, massive and non-fissile in character, with a general absence of laminae. The constituent particles seem to have been deposited at a faster rate than in shales.

Although the bulk of the particles in argillaceous rocks are less than 0·004 millimetres in size, it is common experience to find an ample proportion of silt-sized particles. Most of the rocks contain a mixture of two siliciclastic materials, rock-flour and clay. Many glacial clays and aeolian dusts consist principally of particles of quartz, feldspar and mica; this material is known as rock-flour and, except in its grain size, differs little from the material of sands.

True clays, on the other hand, consist of the minute flaky crystals of minerals produced by chemical weathering of feldspars and other destructible minerals. They differ fundamentally in their composition and properties from coarser deposits. In addition to these materials, chemical precipitates, organic matter and various colloids are commonly found in varying proportions as accessory constituents. Marls usually consist of clay mixed with calcium carbonate, either of organic origin as in the Chalk Marl (Cretaceous) of Europe or in the form of a biochemical precipitate.

COMPOSITION
Rock-flour

The mechanically formed rock-flour which is present in various proportions usually consists of quartz, feldspar, muscovite, and biotite, but granules of other rock-forming minerals may be present. These minerals are reduced to a fine powder either by abrasion during transportation, or by crushing. In addition, minute crystals of stable minerals such as zircon, tourmaline, and rutile are released from igneous rocks by the decay of the less stable minerals in which they are embedded, and ultimately find their way into the argillaceous deposits.

Table 7.1 Chemical analyses of clay minerals.

	1 Kaolinite	*2* Montmorillonite	*3* Illite	*4* Chlorite	*5* Glauconite
SiO_2	45·44	51·14	42·96	26·28	52·64
Al_2O_3	38·52	19·76	28·97	25·20	5·78
Fe_2O_3	0·80	0·83	2·27	—	17·88
FeO	—	—	0·57	8·70	3·85
MgO	0·08	3·22	1·32	26·96	3·43
CaO	0·08	1·62	0·67	0·28	0·12
Na_2O	0·66	0·04	0·13	—	0·18
K_2O	0·14	0·11	7·47	—	7·42
H_2O-	0·60	14·81	3·22	—	2·83
H_2O+	13·60	7·99	6·03	11·70	5·86

1 Roseland, Virginia.
2 Montmorillon, France.
3 Ballater, Scotland.
4 Ducktown, Tennessee.
5 New Zealand.

Figure 7.1 Vermicular kaolinite. The grains are partly stained by carbonaceous matter. Crossed polars, × 150.

Clay minerals

Only a few of the crystalline clay minerals can be considered here (Table 7.1). Of the two-layer types, with a sheet structure composed of alternate layers of silica tetrahedra and alumina octahedra, the **kandite** group is probably the best known. Kandites have an approximate composition of $Al_2O_3.2SiO_2.2H_2O$ in which ferrous and magnesium ions often partially substitute for aluminium ions. Individual members include kaolinite *sensu stricto* (Fig. 7.1), dickite, chamosite and greenalite. Halloysite is a related form.

Of the three-layer types, with sheet structures composed of two layers of silica tetrahedra interleaved with alumina di- and trioctahedra, the best known are the **smectite** and **illite** (hydromica) groups. The smectites include montmorillonite and nontronite and are characterised by an ionic lattice capable of expansion and contraction. This physical change is brought about by the adsorption or loss of water molecules. The illite group by contrast consists of non-expanding minerals which have a complex chemical composition resembling that of muscovite and sericitic mica. Apart from the variety illite, the group contains glauconite (in the strict sense). Closely related in composition are members of the chlorite group in which ferrous ions are prominent. The chlorites have a mixed layer type of structure with alternate two-layer (kandite) and three-layer (smectite) type arrangements of ions. Many clay minerals other than the chlorites appear to be mixed layer type and are commonly referred to as chlorite-illites, kaolinite-illites, etc. instead of devising new specific names for each mixture.

When certain clay minerals, notably montmorillonite, are in equilibrium with the solutions present in a slightly calcareous sediment or soil, they absorb calcium ions, and in this condition the colloidal clay-complex is known as a **calcium clay**. If a solution of sodium chloride is allowed to percolate through a calcium clay, calcium ions are taken into solution, and are replaced by sodium ions, thus forming a **sodium clay**. This phenomenon is known as base exchange. Reaction with acid solutions, such as rain-water containing carbon dioxide, or with acid humus, tends to remove exchangeable bases, and replace them by hydrogen, the result being a **hydrogen clay**. Such clays are commonly developed in sour or acid soils, from which the calcium has been leached by acid rain-water. Other exchangeable bases are magnesium, and, in cultivated land, potassium and ammonium.

From the sedimentary point of view, the importance of base exchange lies principally in the effects of different replaceable bases upon the permeability and degree of flocculation of clays. Hydrogen clays are highly dispersed, and are impermeable. Sodium clays, such as those deposited from salt water, are flocculated and allow a slow passage to gases and liquids. Calcium clays, which may be deposited from hard fresh-water, are strongly flocculated, and are much more readily permeable than either hydrogen or sodium clays.

If any of these clays are deposited from water containing electropositive colloids, such as ferric hydrosol, or certain organic colloids, the charges on the electronegative clay particles are neutralised, and a highly flocculated precipitate is formed.

Accessory constituents

Siderite and calcite occur both as primary precipitates and as early diagenetic concretions and nodules. In the former state it is rare for the two minerals to be intimately associated, because of differing geochemical requirements for their precipitation.

Authigenic pyrite, marcasite and pyrrhotite grains and nodules are frequent constituents, and owe their origin to an *in-situ* reduction of iron salts by organic compounds or bacteria.

Organic matter, in the form of humic colloids, or as solid animal or plant débris, is often deposited in considerable quantity with argillaceous sediments, and gives rise to various diagenetic products. Organic radicals also attach themselves loosely to the clay mineral lattices.

Ferric oxide, usually in a hydrated condition, is another frequent constituent which is often intimately mixed with the clay minerals. Large quantities of red ferric oxide are transported by streams draining regions of lateritic weathering, and find their way into the resulting fine-grained sediments. In exceptional circumstances, aluminium hydroxides may be similarly transported in quantity, and are redeposited to form bauxitic clays.

DIAGENESIS

Chemical and mineralogical changes are liable to begin almost as soon as the mud is deposited. These earliest changes are often made strikingly apparent by an alteration in the colour of the sediment, which in some environments becomes perceptible within a day or two after deposition.

The clay mineral components of argillaceous rocks show evidence of change starting with mild ionic substitution in lattices and sometimes ending with complete conversion, such as montmorillonite into illite (Fig. 7.2). Montmorillonite forms 60 per cent of the clay mineral content in some Gulf Coast sediments at a depth of 1800 metres below the present surface. At 4500 metres depth the montmorillonite content is reduced to 20 per cent, probably due to conversion into illite and mixed-layer minerals. However, the assumption made in these assessments is that the change is entirely diagenetic and unrelated to progressive changes in provenance. Large amounts of interlayer water may be released from the clay minerals if the changes are truly diagenetic, thus creating abnormally high pore pressures in the sediment. Some of these waters may be capable of mobilising oil and bitumens and transporting them into adjacent, more permeable beds.

In muds containing iron pyrites, oxidation of the mineral forms sulphuric acid which then attacks granular kaolinite and forms, in its place, vermicular kaolinite and dickite. Montmorillonite can form from halloysite, and kaolinite form from feldspar, by the action of siliceous waters. Small illites sometimes recrystallise to give large hexagonal grains two millimetres across.

Clay minerals almost certainly migrate in colloidal solution at early compactive stages while porosity is still relatively high. The smectite group minerals are usually most mobile and often replace the calcite in shell débris. The kandites are less mobile except in acid water, plant-rich environments.

The muds of temperate and tropical rivers, lakes, estuaries and seas usually contain an appreciable amount of organic matter, which supports an active population of micro-organisms, conspicuous amongst which are the bacteria. The surface layer of mud, having access to the dissolved air of the overlying water, tends to remain in an oxidised condition, and supports a population of aerobic bacteria. The iron in this layer is usually in the form of limonite,

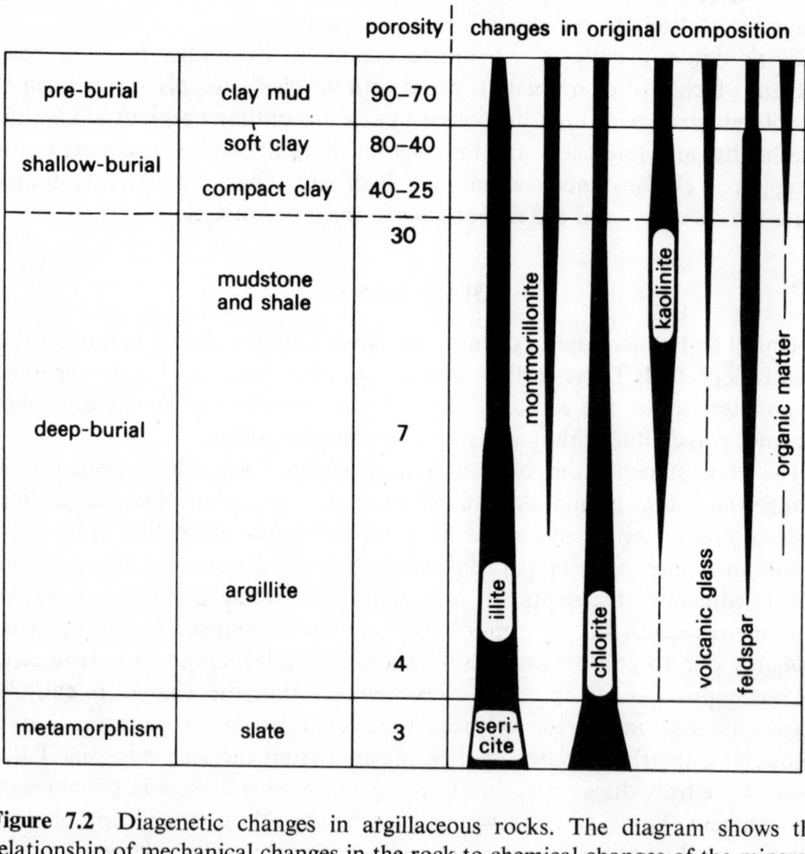

Figure 7.2 Diagenetic changes in argillaceous rocks. The diagram shows the relationship of mechanical changes in the rock to chemical changes of the minerals within the rock.

which gives a brown or yellowish colour to the sediment. At a slight depth, sometimes merely a fraction of a centimetre, circulation of oxidising solutions is sufficiently restricted to allow reducing conditions to prevail, the reducing environment being intensified by an active population of anaerobic bacteria. The sulphates are reduced, and organic substances such as proteins and carbohydrates are broken down, with the production of such gases as hydrogen sulphide and methane. Iron compounds are reduced and converted to sulphide, colloidal $FeS.H_2O$ being first formed, giving a dark, often black, colour to the deposit. This becomes converted to the black disulphide melnikovite, which at a late stage of diagenesis (probably during consolidation), slowly changes to pyrite.

Compaction

When newly deposited from suspension in water, argillaceous sediments remain for some time in a fluid condition; the component particles, which are commonly minute flaky crystals, are each encased in a sheath of adsorbed water, and at this stage are not in physical contact with each other. The sediment in bulk contains only from 10–30 per cent of solid matter by volume, and may be said to have a porosity of 70–90 per cent, much of the space being occupied by mechanically enmeshed water. Muds containing much finely divided organic matter may have initial porosities exceeding 90 per cent, and flocculated deposits in general have a greater initial porosity than those which settle in a dispersed condition (Fig. 7.3).

The clay particles begin to pack together to form a less open network, especially if more sediment continues to accumulate above. Free water is expelled as the larger interspaces become reduced, but the sediment remains fluid until the pore space has diminished to 75 per cent or less.

Under an increasing load of strata, further expulsion of free water takes place, and the aqueous sheaths of adjacent grains are brought nearer together until they are in close contact. The expulsion of water becomes increasingly difficult as the pore space diminishes, since the frictional resistance offered to the passage of water becomes correspondingly greater. The weight of overlying strata is now principally carried by the aqueous envelopes of the particles, but with still increasing loads the solid grains break through the adsorbed films at the points of greatest pressure, and the mineral particles themselves come into contact. An aqueous sheath still encloses the remainder of the surface of each grain. It is believed that the porosity at this stage is usually between 30 and 35 per cent.

So far, the compaction of the sediment has been effected principally by expulsion of free water, accompanied by mechanical adjustments of the solid particles. A considerable proportion of the free water has now been removed, and further consolidation involves expulsion of part of the adsorbed water, deformation of the mineral particles themselves, since they are now in physical contact, and precipitation of chemical cements.

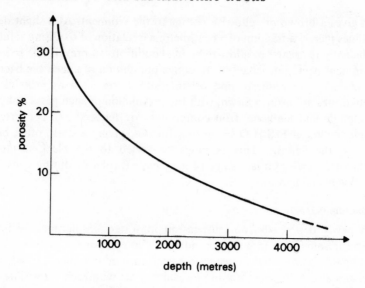

Figure 7.3 Porosity in shales. The graph illustrates the progressive decrease in porosity with increasing depth of burial for some Liassic clays of north-west Germany.

The physical strength of clays is basically a function of decrease in pore space, this usually being a consequence of overburden pressure exerted by the sediment itself. In deep water some pelagic clays are often stronger than would be anticipated as a consequence of simple pressure and are referred to as being **overconsolidated**. In all probability the overconsolidation of these sediments reflects slow rates of deposition and enhanced rates of chemical cementation. Strictly, the soils mechanics term 'overconsolidated' should only be applied to sediment which has been subject to overburden pressures greater than those presently effective at the depositional surface. But, the usage of the term has been extended so as to include pore space reduction effected by desiccation, cementation, unusually strong interparticle forces and other factors. For example, the upper layers of alluvial and supratidal mud during dry climatic phases become much more rigid than the immediately underlying sediment and can be said to be overconsolidated. That is, they have hardened at a faster rate than would have happened if the sediment had been subject to 'normal' compactive processes. Subsequent deposition can then lead to the incorporation of these overconsolidated layers into the pile of softer sediments. These layers are best seen in recent partially lithified sediments; they are very difficult to detect in a fully lithified succession. On the other hand, induration of this kind is often reversible, and the sediment breaks down again if allowed to remain sufficiently long in contact with an excess of water.

Rapidly deposited shallow water clays may be less strong near to the surface than anticipated and may be termed **underconsolidated.** In these cases the overburden stress is largely borne by the pore water rather than the solid particles. The marine clays being deposited adjacent to the actively prograding Balize delta lobe of the Mississippi River are generally underconsolidated. During phases of very fast deposition and loading these clays often distort and flow. Fracturing also occurs though the planes of fracture tend to reseal without leaving slickensided surfaces.

Rheotropy, thixotropy and dilatancy

The role played by pore water in the flow and consolidation of mud has attracted much attention though its effects in many instances are often difficult to confirm. The science of the flow of materials is called **rheotropy** and the study of natural muds is simply one small field of endeavour. None the less it is known from laboratory evidence that if a firm, but unconsolidated, mud is subject to vibration it will either become partially mobile or completely mobile. These two states are referred to as **false body** and **sol** respectively and seem to be created by changes in the packing of the grains. This results in pore water being released and immediately makes the mud more mobile. In the true sol state the mud flows under the slightest shearing force. When the vibration ceases the liquified mud once more becomes firm and readsorbs most of the water released earlier; this is called the **gel** state. In this state flow will only occur when vibration or a shearing force is reapplied and reaches a minimum critical value. This value is called the **yield point** and is naturally dependent on factors such as the mineralogy of the mud and its grain packing. Smectite-rich muds (bentonites) have a low yield point, rapidly becoming fluid when vibrated and equally rapidly returning to a gel or firm state when the vibration ceases. Illite-rich muds flow less easily and have a higher yield point.

Rheotropic changes can also be accelerated in the presence of suitable electrolytes and it is known that muds, in general, tend to have a lower yield point in marine waters than in fresh. Clays raised above sea level and leached of sea salts are known to increase in yield value and decrease in plasticity by as much as 60 per cent.

Sediments which have low yield points are usually called highly thixotropic. Most clays are of this type. The fact that many thick clay horizons (Mesozoic Weald Clay, Gault, Fuller's Earths of southern Britain) show poor or little bedding may be one expression of rheotropic behaviour, the original bedding planes being destroyed by penecontemporaneous mobilisation.

Whereas highly thixotropic muds increase in mobility with increase in shear there are others which decrease in mobility with increase in shear. The latter are said to have **dilatant** characteristics. Some soft fluviatile muds when compressed, vibrated or simply walked over, become relatively firm but rapidly return to a soft state when the stress has been removed. Dilatancy, as applied to muds, has not been recognised as a significant process in fossil sedimentation. This is probably due to lack of suitably preserved criteria in the rocks for its recognition.

CLAY MINERALS AND ENVIRONMENT

Although clays, mudstones and shales consist of a variable admixture of clay minerals and other clastic particles it is appropriate to consider at this point the distribution of clay minerals in sediments.

The majority of clay minerals which eventually find their way into the complex of water-covered environments are eroded from a mixed terrain of exposed rocks. In these areas of subaerial weathering and soil formation the physical and chemical environments are probably as important as the composition of the parent rocks in determining the type of clay mineral produced. The formation of kandite group minerals in soils is favoured if rainfall exceeds evaporation and leaching is intense. Strong leaching removes Ca, Mg, Na and K ions if they should be present in the parent rocks and stabilises any iron by converting it into oxide or sulphide form. Excessive silica ions also need to be removed so as to maintain a high Al:Si ratio in the soils. Under these conditions the kandites will form irrespective of whether the parent rocks are granites, gabbros or volcanic ashes.

The smectites are generally formed under conditions which are the reverse of those needed for kandite production. Evaporation should exceed precipitation and leaching processes should be negligible and alkaline conditions prevail so that a low Al:Si ratio is maintained.

Non-acid, potassium-rich conditions favour illite creation. Rainfall and consequent leaching should be only moderate and intermittent. A lack of sufficient potassium in the soils usually means that a degraded variety of potash-deficient illite forms but conversion to normal illite is soon accomplished when the illite is transported into an alkaline marine basin.

Some petrologists consider that clay minerals alter little, if at all, during fluviatile transport but become very susceptible to change once they reach open bodies of water such as large lakes, seas and oceans. It has been suggested that large changes in composition occur in fluviatile clay deposited in saline waters, with the preferential formation of smectites, illites and chlorites and loss of kandites. If the clay is not buried quickly and the ionic composition of the water is appropriate then the changes may occur relatively rapidly.

Figure 7.4 illustrates changes in clay mineralogy detected in bottom samples over a distance of 160 km extending from the Guadalupe River in Texas via Aransas Bay into the Gulf of Mexico. The increase in salinity and chlorinity of the waters is believed to be reflected by the increasing conversion of montmorillonite into chlorite.

Other petrologists are very sceptical of drastic changes in clay minerals during and after deposition. They conclude that most clay minerals laid down are detrital in origin and strongly reflect the character of the rocks in the source area. Authigenic glauconite is an important exception. In their view any of the major clay minerals can occur in abundance in any of the major depositional environments and there is no consistent genetic association between specific environments and specific clay minerals. The only modifications admitted by these workers are secondary cation adsorption and loss (base exchange). These changes do not alter the basic clay mineral lattices.

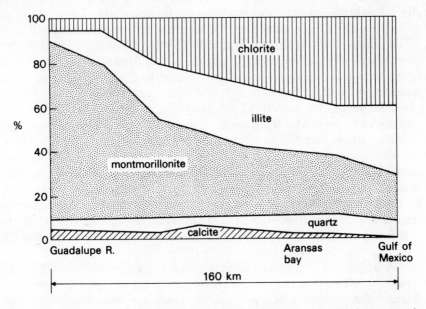

Figure 7.4 Clay mineral distribution in modern sediment near Rockport, south Texas. The diagram shows the mineral distribution in the less than one micron size fraction of sediment being transported and deposited in the lower reaches of the river and adjacent sea. The marine influence increases progressively from the mouth of Guadalupe River, via Aransas Bay, which is protected by a coastal barrier, into the open Gulf.

The clay minerals laid down at the mouths of the Mississippi delta do not show any significant differences from those being carried downstream in the upper reaches. Only along the outermost seaward edge of the delta is there any suggestion that conversion of river-borne montmorillonite to illite is occurring. Even here, the evidence is indirect and is simply based on the fact that illite is more common in these muds than in those nearer to the delta proper. A similar indirect line of reasoning is used for the clay mineral distribution off the Californian coast. Adjacent to the land, especially near to the Colorado River and San Diego, kaolinite, montmorillonite and illite are about equally abundant in the muds. In the off-shore basins illite increases in proportion (to 40 per cent) and kaolinite decreases (to 20 per cent), hence it is assumed that the kaolinite is being converted into illite. There is no doubt that sedimentation in the off-shore basins is slower, thus providing more time for alteration to take effect and there is little doubt that the abundance of calcareous organisms in the off-shore muds is not conducive to kaolinite stability. But whether this means an extensive structural and chemical conversion of the kaolinite into illite remains doubtful. The variations may be accounted for by the reworking of relict clay materials on the sea bottom or the differential distribution of incoming clastic clay grains.

In the Gulf of Paria off the north coast of South America the distribution of clay minerals shows the influence of flocculation processes. These processes are undoubtedly very important in clay deposition. Upon contact with marine water, clay particles suspended in fresh water tend to flocculate and form larger composite particles. In the Gulf, illite and kaolinite appear to be flocculating quickly usually as relatively coarse particles which are then rapidly deposited. Montmorillonite is flocculating more slowly as small particles which take longer to be deposited. Consequently, there is a progressive sorting of the clay particles such that the illite- and kaolinite-rich zones correspond with the fresh and brackish water marginal deltaic areas and the montmorillonite-rich zones with the more open marine areas.

Changes in the climate of a source area may be ultimately reflected in the clastic clay mineral content of derived sediments, though the changes in the sediments are likely to be gradual and spread over at least a moderate thickness of strata. On the occasions when a sharp change in clay mineral composition is detected at a precise level within a sequence other considerations arise. For instance, in conformable beds the marked change may represent a hiatus of considerable duration, during which a significant change in the nature of the provenance or a total change of provenance occurred.

MARINE DEPOSITS

The influence of marine processes on mud deposition is felt from immediately adjacent to coastlines, outwards across continental shelves, and into the deeper oceanic basins. The distribution of the muds follows no regular pattern being controlled by bottom topography, the pattern of marine currents and the location of the clay sources. Off the mouths of major deltas, such as the Mississippi and Orinoco, the clays laid down in shallow marine waters are little different in lithology from those deposited well away from any deltaic influence.

In shallow shelf areas of mud and silt accretion the rate of accumulation may be so slow that colonies of various kinds of marine organisms have time to rework thoroughly the sediments. A complex mottled or bioturbate texture is produced.

Blue or slate-gray mud of terrestrial origin is widespread in modern seas and oceans. Green muds, such as those in the basins off-shore from California, seem to be much more common than originally realised. The green colour is caused by green illite and montmorillonite; chlorite and fine-grained glauconite are very subordinate in amount. A high organic matter content (about 7 per cent) is a typical feature of green basinal muds. Foraminiferal tests, which show little evidence of solution, often form an important proportion of the deposits. Green shales and mudstones are common in the Lower Palaeozoic rocks of Wales.

Marine argillaceous sediments derived from lateritic land areas tend to contain conspicuous quantities of red ferric oxide. The red muds of the western Atlantic are believed to owe their ferric oxide pigment to lateritic material transported by the Amazon and Orinoco Rivers.

Black muds and shales

In a few parts of modern seas, black muds containing considerable quantities of undecomposed organic matter, and large concentrations of carbon dioxide together with varying amounts of hydrogen sulphide, are developing where there is a strong oxygen deficiency in the water immediately overlying the sea floor. Depth of water, in itself, has little or no controlling influence upon the formation of such sediments, the principal requirement being lack of circulation. Thus black muds may be deposited at almost any depth, varying from the shallowest pools down to profound depressions such as that of the Black Sea Basin and deeper Norwegian fiords, provided that the water in contact with the sea floor is foul and stagnant. The oxygen deficiency in such an environment inhibits benthonic animal life, and any remains which may be found are those of planktonic or nektonic creatures, whose dead bodies have sunk from the better-aerated surface waters. The absence of benthonic life from such an environment appears to be primarily due to this lack of an adequate supply of oxygen, rather than to the toxic effects of such substances as hydrogen sulphide or carbon dioxide.

Examples of ancient black shales which were probably formed under similar conditions are abundant. The black graptolitic shales of the Lower Palaeozoic are pigmented partly by finely disseminated iron sulphide, and partly by carbon, which amounts to as much as 6 per cent in some beds. Much smaller quantities will suffice to give a sooty black colour to the rock, and many analyses of black shales show about $0 \cdot 5$ per cent of carbon, and between 1 and 2 per cent of iron sulphide. The absence or rarity of any remains of bottom-living organisms in these deposits, whilst planktonic forms are often abundantly preserved, suggests that whereas the surface waters were wholesome, the layer in contact with the sea bed was unfavourable to animal life.

Black or dark-coloured shales of basinal facies are found at various horizons in the Palaeozoic and Mesozoic, and the fauna is almost invariably restricted to planktonic forms. In north-west Europe the goniatite shales of the Devonian and Carboniferous, and some of the dark bituminous shales of the Lower Jurassic present a remarkable uniformity, both in their lithology and in the bionomic character of their faunas. The rocks are often pyritic, and not infrequently slightly oily. Their tendency to weather into thin laminae gives rise to the term 'paper shales'. Bottom-living forms are characteristically absent, and the fauna principally consists of cephalopods, together with lamellibranchs such as *Posidonia*, which is believed to be a byssus-bearing form, living attached to free-swimming cephalopods or to floating seaweed.

Where black shales commonly differ from other shales is in containing higher concentrations of minor elements such as uranium, arsenic, copper, molybdenum, lead, vanadium and zinc. Clearly the geochemical circum-

Table 7.2 Chemical analyses of argillaceous rocks.

	1	2	3	4	5	6	7	8
SiO_2	49·83	29·82	43·05	40·40	30·11	53·71	71·39	34·60
Al_2O_3	15·30	12·16	18·52	33·50	42·29	13·73	18·05	13·90
TiO_2	1·44	0·48	0·78	7·00	5·83	—	—	0·55
Fe_2O_3	6·73	1·73	0·68	0·50	2·81	2·78	1·62	—
FeO	4·97	7·12	13·16	—	0·72	1·86	0·32	1·00
MgO	3·72	5·60	3·34	—	0·82	4·92	1·25	1·10
CaO	1·32	10·38	1·46	1·40	0·13	6·38	0·07	12·45
MnO	0·17	0·31	0·32	—	tr	—	—	0·14
Na_2O	1·91	0·34	0·66	0·04	0·36	0·53	0·02	0·14
K_2O	2·66	1·74	1·43	0·25	0·26	3·84	3·07	1·63
H_2O-	0·43	2·08	1·05	—	—	0·90	0·66	—
H_2O+	3·92	4·06	6·03	—	16·53	3·61	—	4·84
P_2O_5	0·22	0·92	0·33	1·05	tr	—	—	0·14
CO_2	0·13	10·57	6·18	—	—	—	—	10·80
Organic C	0·66	10·93	2·72	—	—	—	0·04	11·51
S	2·44	2·44	0·12	0·07	—	—	—	—
Cu	10·11	—	—	—	—	—	—	0·02
								6·45 FeS_2
								0·18 BaO

1 Nonesuch Shale, Precambrian, Michigan.
2 Pot Clay Marine Band, Carboniferous, Derbyshire.
3 Non-marine mussel band, Carboniferous, Derbyshire.
4 Tonstein, Carboniferous, Fife.
5 Bauxite, Tertiary, Co. Antrim.
6 Boulder clay, Pleistocene, Illinois.
7 Clay, Cretaceous, Illinois.
8 Marl Slate, Permian, Durham.

stances in these specialised environments favours the precipitation of these elements. Adsorption by organic complexes and clay minerals is probably an important retention process. Syngenetic metallic sulphides, sometimes in economic quantities, are also concentrated in the shales. The Permian Kupferschiefer–Marl Slate extending from Poland to northern England has yielded locally up to 3·6 per cent copper and 1 per cent zinc from sulphides. Up to 10 per cent copper, mainly as sulphide, has been recorded in the two black shale horizons of the late Keeweenawan (Precambrian) Nonesuch Shale of Michigan (Table 7.2).

Brown (red) clay

By far the most widely spread and most characteristic of all the abyssal deposits is that to which the name of brown clay is applied. It is formed to some extent by the disintegration of rocks and minerals *in situ*, but to a greater extent by fine terrestrial detritus being transported from shelf areas and redistributed by deep-sea currents.

The older name, red clay, is now considered obsolete because the deposit typically has a chocolate-brown colour. It occurs in all the deepest parts of the oceans, and in the Pacific it is estimated to cover an area of nearly 70 000 000 square kilometres; it is also abundant in the Indian Ocean and in the deeper parts of the Atlantic, and, in fact, it is found in almost all regions where the depth of the water exceeds 4·5 kilometres. It is plastic when wet, but dries to a hard mass, showing many of the characteristics of clay. Its carbonate of lime content is low, not exceeding 4 per cent; but siliceous organisms are often abundant, and brown clay graduates into radiolarian ooze. Sponge-spicules are generally present, and sometimes diatoms. The mineral components of the brown clay show a good deal of variation according to the locality: in the south Pacific the most abundant constituents are minute fragments of basic volcanic rocks; in the north Pacific pumice prevails; while in the south Atlantic quartz is dominant, with some feldspar and clay minerals. These particles are partly derived from the decomposition and decay of fragments ejected by volcanic eruptions, some of which may have been submarine. The fact must be strongly emphasised that brown clay has no definite composition, but varies according to the source from which the material is derived; it is believed that the eruption of Krakatoa in 1883 strongly influenced the composition of the abyssal brown clay deposits of the Indian Ocean. In some parts of the Atlantic are found grains of sand which must have been carried by the Harmattan winds from the deserts of Africa.

Another substance common in samples of brown clay is phillipsite, which is often very prominent and together with related zeolites can form up to 50 per cent of the sediment.

Diagenetic rounded and botryoidal ferro-manganese nodules are abundantly formed just at and below the surface of brown clays. They are scattered over the floor of oceans, and it has been estimated that at least 10 per cent of the Pacific area is covered by them. The nodules vary in size from pellets a few microns in diameter to slabs several metres wide. Laminar internal structure is usual which may be connected with the activities of encrusting arenaceous foraminifera and fungi around the periphery of the growing nodules.

The nodules are often associated with vast numbers of sharks' teeth and the ear-bones of whales. The abundance of these organic remains indicates that at the greatest depths the deposits form with extraordinary slowness, since remains of extinct species are often dredged up along with those of existing forms, showing that since Tertiary times the thickness of brown clay formed has been insufficient to bury them completely. Apart from these larger constituents, the brown clay is summed up as a decomposition product of aluminous silicates derived partly from the rock fragments spread over the oceans by various subaerial, submarine and volcanic agencies, and partly from 'weathering' in place of the rocks forming the sea floor: from one point of view it can be regarded as a residual deposit from which most of the carbonate of lime and silica have been removed by solution.

Marine marls and marlstones

Marine marls, marlstones and calcareous shales are, in some instances, impure calcareous organic deposits, such as the Cretaceous Chalk Marl of south-eastern England, which owes much of its calcareous nature to an abundance of coccolith plates and fragments of these. In other cases, as in some of the marly limestones of the Lias, a certain amount of precipitation of calcium carbonate appears to have taken place at the surface of the mud during or shortly after deposition.

Bentonite

Bentonite is an argillaceous rock of peculiar character, known principally from the Cretaceous and Ordovician of North America. Unweathered material is light green or pale greenish-yellow in colour, and has a fracture rather like that of hard wax. A property of this material is its marked capacity for base-exchange.

Bentonite consists essentially of montmorillonite, or closely allied clay minerals, in extremely minute crystals, giving a very large total surface area. Most samples also contain small particles of orthoclase, plagioclase, and biotite, and some of the accessory minerals of igneous rocks. Thin sections show relics of the characteristic structure of pulverised pumice. Bentonite is believed to be formed by the devitrification of volcanic ashes in which glass is the principal constituent. The Cretaceous bentonites of Arkansas are interbedded with volcanic tuff made up of angular fragments of trachyte, and in this case there is no doubt that we are dealing with an altered trachytic glass, which was originally deposited as a pumice-ash. Other bentonites, such as those south of the Black Hills of South Dakota, appear to be derived from andesitic ashes.

In the alteration of volcanic glass to bentonite, devitrification is accompanied by certain chemical changes. The alkalis, which are important constituents in glasses of feldspathic composition, are almost completely removed, and a considerable proportion of silica is also lost. The principal additions are water of hydration, magnesium, and iron.

Although bentonite deposits are usually unfossiliferous, they are almost invariably interstratified with marine sediments, and there can be little doubt that they were laid down in salt water. Individual beds are usually quite thin, varying from about three centimetres to a metre in thickness, but in spite of this, they remain surprisingly uniform in character over enormous areas. The principal deposits are found in the Cretaceous of western North America, and in the Ordovician of the Appalachian region. It is highly probable that some of the rocks described as fuller's earths in the Silurian, Jurassic, and Cretaceous of England and Wales should really be regarded as bentonites; the Lower Cretaceous fuller's earth of southern England, for example. Bentonite lenses and seams are also now recognised in Carboniferous marine and non-marine successions of northern England and Scotland. Certain of these deposits may be regarded as water-laid tuffs which decomposed *in situ*; in other cases there is reason for supposing that the material may have been deposited as a montmorillonite clay, derived from the erosion of weathered volcanic rocks laid down elsewhere.

FRESH-WATER AND TERRESTRIAL DEPOSITS

Clays of glacial lakes

Clays of glacial lakes consist principally of material removed mechanically from the parent rock and crushed to a fine powder or rock-flour. In Sweden, where considerable glacial erosion of crystalline rocks has taken place in Pleistocene and recent times, rock-flour is the principal constituent of many lacustrine deposits.

Argillaceous deposits laid down from suspension in the cold waters of glacial lakes commonly show well-marked lamination, as in the **varve clays**. Each double layer or varve in these deposits is believed to represent the sediment accumulated during one year, and observation on the rate of deposition in glacial lakes supports this belief. Sediment is carried into the lake only during the warm months when melting is in progress. The coarse grains sink with comparative rapidity, but the finer material sinks extremely slowly, partly because no flocculation takes place in pure glacial water, and partly because the viscosity of water is much increased near freezing point. It is believed that sedimentation of fine clastic particles continues throughout the winter beneath the surface ice of lakes which do not freeze solid. The coarse (usually silty) deposits of the summer layer thus merge upwards into the extremely fine-grained winter layer, to be succeeded abruptly by the coarser material of the next summer layer.

Clays and muds of non-glacial lakes

Clays and muds of non-glacial lakes usually contain a high proportion of true clay minerals, together with organic matter. Carbonaceous remains of plants are often abundant in the muds deposited in the marginal parts of lakes, whilst in the deeper central parts remains of planktonic algae may form an appreciable part of the sediments. Algae were probably the mother substance of the kerogenous constituents of many finely laminated, lacustrine oil shales, such as those in the Green River formation (Eocene) of the United States of America and in the Oil-Shale Group (Carboniferous) of Scotland. The muds of some lakes are highly siliceous owing to the abundance of diatomaceous material. Such diatomaceous deposits are not uncommon in the post-glacial lakes of Britain, especially where the supply of terrestrial debris was small.

Fine laminations are characteristic of quieter, often deeper water, areas in lakes. The laminae frequently represent annual or seasonal deposition, but may also represent spasmodic and rapid, non-seasonal influxes of sediment, possibly by small-scale turbidity currents.

While most clays probably contain a mixture of clay minerals, deposits consisting principally of kaolinite are known, as for example in the ballclays of south-western England. The Oligocene lake deposits of Bovey Tracey and Petrockstowe consist of the débris of kaolinised country rocks and plant débris, washed down by streams draining the highlands of Dartmoor. The quartz was deposited in extensive sandy deltas at the margins of the lake, the finer clay and plants being drifted further out to form beds of white pottery clay and lignite.

Lacustrine clays of bauxitic composition (i.e. containing appreciable quantities of hydrated aluminium oxides) have been described from the Carboniferous rocks of Ayrshire in Scotland, where they appear to have been formed by the redeposition of bauxite eroded from lateritised basalt. Aluminium hydroxides are infrequent constituents of transported clays, and the great majority of bauxitic clays are residual deposits.

Tonsteins

These are light-brown, kaolinite-rich compact rocks associated with coal-bearing strata. The beds average five centimetres thick (two metres maximum)

Figure 7.5 Tonstein, Coal Measures, north England. Colourless crystals and vermicules (internal structure emphasised) of kaolinite set in a matrix of finer-grained carbonaceous matter and clay minerals, the latter being predominantly organic-stained kaolinite. Crossed polars, × 100.

and usually rest directly on top of a coal seam or are intimately intercalated within the coal. Many have a large lateral extension and thus form useful marker horizons for correlation purposes. In thin-section the dominant kaolinite can be seen to exist in the form of large vermicules or lamellar grains, microcrystalline spherules, or as a cryptocrystalline groundmass (Fig. 7.5 and Table 7.2). Illite and mixed illite–kaolinite grains also occur together with a minor sprinkling of other detrital material.

The origin of these rocks is obscure. Some authors suggest a detrital origin for the kaolinite, probably by the erosion of a deeply weathered granitoid terrain. Others prefer an in-place alteration of volcanic ash, this accounting for the wide areal distribution. This style of alteration would be favoured by an acid environment with pH 3·5–7·0. Degradation would occur under the influence of humic acids concentrated at plant-rich levels. Conversion might be assisted biochemically by the activities of micro-organisms, such as bacteria and algae, under conditions of fluctuating water table.

Tonsteins have to be distinguished in origin from **fragmental-clay rocks,** which are also associated with coal-bearing sequences and contain between 50–90 per cent kaolinite, most of which is authigenic. The distinction rests in the flow-brecciated internal structure of the fragmental-clay rocks, which is believed to be a post-depositional compaction effect.

Fresh-water marls and marlstones

Fresh-water marls and marlstones are sediments intermediate between the clays and limestones, and include every gradation between calcareous clays and muddy limestones. The amount of calcareous matter intermingled with the clay and silt particles should range between 40 and 60 per cent. The degree of lithification is very variable and not necessarily a function of age, so the point at which it is preferable to use marlstone instead of marl is quite subjective. Many of the red Permo-Trias examples in the northern hemisphere are still comparatively soft and easily weathered so the name marl is appropriate. The beds are frequently massive and lenticular and appear to have been deposited rapidly in shallow lakes, possibly by wind activity. In contrast, a large proportion of the lacustrine Eocene Green River formation deposits are dense, hard rocks best described as marlstones

Marls are sometimes formed in lakes by the activities of plants, especially calcareous algae. The Green River beds are a prime example. If the water is saturated with respect to calcium carbonate, extraction of carbon dioxide by the algae results in the formation of precipitate, which may be deposited as a carbonate mud. The alga *Chara* deposits low magnesian calcite within its tissues, and thus builds up a skeleton which is almost wholly calcareous. In most lake marls, the calcium carbonate formed in this way is mixed with argillaceous sediment, other plant debris and calcareous shells. Fresh-water marls containing the remains of *Chara*, together with *Limnaea*, *Viviparus* and *Planorbis* are to be found in the Bembridge and Headon Beds (Tertiary) of southern England.

Loess

Fine-grained aeolian deposits of purely detrital origin are usually grouped together under the name of loess, and are extensively developed in China, North America and northern Europe. In constitution they differ strongly from most water-transported muds and clays, since they consist of chips of rock-forming minerals, and the true clay minerals are either absent or present in subordinate quantities only. The particles are small and sharply angular and probably the product of mechanical fracture in the source area, either by the grinding action of ice or by intermittently active soil-producing processes. Much of the material in the vast Pleistocene deposits of the northern hemisphere shows a very close resemblance to fine-grained glacial débris, and appears to have been picked up from the outwash of glaciers and ice-sheets. On the other hand, there is considerable evidence that the finest detritus formed by wind abrasion in large deserts, such as those of central Asia and western America, is blown out of the desert areas, and accumulates in vegetated regions, and especially in grassland such as the steppes or the prairies, to form a loess-like deposit.

The loess of China and of Europe is a fine calcareous silt or clay which is entirely unstratified and very uniform in texture. It is quite soft, and crumbles between the fingers. It resists denudation, however, in an extraordinary manner, probably on account of its homogeneity, and often stands up as vertical walls tens of metres high. This property is probably assisted by the presence of numerous fine tubes arranged vertically and lined with calcium carbonate; they are supposed to have been formed in the first place by fibrous rootlets.

The loess of central Europe is, for the most part aeolian, formed by redistribution of fine glacial mud originally laid down in water, and, after drying, carried by the wind often to considerable heights. A part, however, of the so-called loess of northern France, Belgium and south-eastern England is loessic material reworked by solifluction and rain downwash processes, and is preferably referred to as **brickearth**. **Adobe** is a deposit similar to loess and largely developed in the Mississippi valley and elsewhere in the United States.

SOILS

In most parts of the world, the purely mineral residues of weathered rocks become mixed with organic substances, principally derived from the débris of plants. The activities of animals and plants, and the reactions of organic solutions, modify the structure and composition of this mixture to produce soils. The nature of the soil at any particular locality is principally controlled by the nature of the underlying rock, the climate, and the topography. In the early stages of weathering the influence of the parent rock is naturally very strong and very largely controls the nature of the soil in regions of immature soils such as Great Britain.

Climate always exerts an important control upon the relative effectiveness of the various weathering processes, upon the biological factors in soil formation, and especially upon the distribution of soluble substances in the soil. It is found that in regions of maturely formed soils, such as eastern Europe or the central United States, the influence of the parent rock is minimised, and uniform soil conditions tend to be developed over wide areas within definite climatic belts. The influence of topography is felt in several ways: in regions of appreciable relief, erosion of the upper layers of partially developed soils prevents the soil from reaching a mature condition, so that there is always a considerable proportion of incompletely weathered material, and the influence of the parent rock is strongly evident. In peneplaned regions, on the other hand, mature and uniform soils develop over wide areas.

Soil-forming processes

The process of weathering, which provides the mineral basis of soils, is the first and perhaps the most important stage in soil formation. From the pedological (pedology, the study of soils) point of view, however, the residual substances formed by this process can be regarded as no more than raw materials, for a fully developed soil possesses special peculiarities in structure and physical properties which distinguish it from a purely residual deposit. The most important of these properties depend upon the presence of a complicated system of colloidal substances, both organic and inorganic, by virtue of which the soil is enabled to retain considerable resources of moisture and exchangeable bases within itself.

Organic matter, derived mainly from plants but also partly from animals, is added at the surface, and mineral substances are supplied from below by subsurface weathering; the resulting mixture of organic and residual matter is subjected to certain further changes which lead to the construction of soils as opposed to undifferentiated mineral residues.

The mantle of soil overlying its parent rock normally consists of several layers or horizons, differing from each other in structure and composition. The succession of these differentiated horizons from the unweathered rock up to the surface is known as the **soil profile**. The differences to be observed between successive horizons in the soil profile do not result solely from an increased maturity of weathering as the surface is approached, but also from the selective redistribution of mineral and organic matter by water migrating through the soil.

In soils with normal drainage, water passes downwards during rainy periods, and upwards in times of drought, when evaporation from the surface exceeds the rainfall. In humid climates such as that of the British Isles, the movement is predominantly downwards, so that soluble salts, colloids, and small mineral particles tend to be removed from the surface layer, known as the A horizon, and transported into the underlying layers of the soil. This process is known as **eluviation**. The soluble salts are completely washed away into the ground-water, but certain of the colloids, especially those consisting of minute mineral particles, are largely redeposited within the soil at some lower level, which is known as an **illuvial horizon**. Eluviation of this kind leads to the formation of heavy clay subsoils with sandy or loamy topsoils. The operation of this process depends upon the dispersion of soil-colloids in the eluviated layer, and is hindered in soils where surface conditions are not favourable to dispersion. Thus the presence of exchangeable calcium, by keeping the clay-colloids in a flocculated condition, prevents their being dispersed into a mobile form, and eluviation can only become effective after the calcium has been leached away from the sphere of activity.

Eluviation of soluble substances in a moist climate leads to the removal of these materials to the ground-water level, and the upper layers of the soil become leached. Calcium carbonate and salts of sodium and potassium tend especially to be removed in this way. This process is called podzolisation.

In arid and semi-arid climates, especially if the level of ground-water is not far below the surface, strong evaporation causes an upward migration of soil-moisture, and the process of eluviation works upwards instead of downwards. As a result of this, we find in some regions that the more soluble salts, instead of being leached away, become concentrated at the surface, giving saline or alkaline soils. There is also a tendency for calcium carbonate, calcium sulphate, alkali silicates, and iron compounds to migrate towards the surface, where they form hard cementing crusts of limestone, gypsum, silica, or ferric oxide. These duricrusts can be up to several metres thick. The limestone varieties are commonly referred to as caliche or calcrete, though not all these varieties have the same origin. Some also originate by downwards seeping water in terrestrial and shoreline environments. In less extreme situations, the salts may be precipitated as grain coatings, veins, concretions and thin layers within the upper part of the soils. Soils enriched in this fashion with carbonates are called pedocals. In climates with seasons of rain and drought, downward and upward eluviation alternate. Since some constituents are irreversibly precipitated, whilst others migrate up and down with alternating movements of the soil-water, rather complicated and highly individualised soils may be produced.

When the water table is high and the lower soil layers saturated, so that a reducing environment exists, the process known as gleying occurs. This creates a mottled brown, olive-gray and grayish-blue coloration, the soil becoming admixed with varying amounts of amorphous organic matter, ferrous iron complexes and salts.

Podzolic soils

These are yellow, brown and pale-gray acid soils in which alkalies and alkaline earths have been removed by intense leaching. Their total thickness is usually less than a metre. Together with closely related types, they cover enormous areas in Russia, Siberia, and northern Europe, and also spread over a considerable area in Canada and the north-eastern United States. The podzols thus occupy a humid climatic belt lying to the south of the tundra.

In the development of a typical podzol, the surface layers are leached of all soluble substances such as calcium carbonate, and the silicates are hydrolysed to form colloidal-clay minerals, ferric hydroxide, and various soluble salts. Intense leaching removes all but the original quartz, so that a bleached sandy layer is formed, which near the surface is mixed with black peaty material. A considerable quantity of humus is usually carried downwards into the subsoil with the other products of leaching. Clay, ferric oxide, and humus tend to become concentrated in the subsoil, and can form hardpans. This layer may become so impervious to water that the course of soil development is altered.

We thus have three horizons in the profile of a typical podzol:

A   Dark peaty surface layer, underlain by bleached quartz-sand (eluviated horizon).
B   Subsoil enriched in humus, clay, and ferric oxide (illuvial horizon).
C   Disintegrated parent rock.

Terra rossa is a particular variety of podzolic soil characterised by a distinct red coloration. It covers the surface of large areas in south-eastern Europe on the borders of the Adriatic Sea, where the underlying rocks are predominantly limestones.

The material itself consists of clay, with small quantities of ferric oxide, and, in the upper layers, a little light-coloured humus; calcium carbonate is usually present only as solid lumps of undissolved limestone. Although the limestones of this region are often white, the fact that they contain traces of iron compounds can be shown by dissolving a large sample in weak acid. The residue thus obtained is usually not red, and the characteristic colour of terra rossa appears to be acquired as a result of changes taking place within the residual material after its release from the limestone. Residual clays of this kind are found overlying limestones in every part of the world where the climatic and topographical conditions are suitable for their accumulation, but are specially characteristic of Mediterranean temperate regions. In regions of heavy rainfall, terra rossa does not accumulate *in situ*, because the residue is washed away as soon as it is formed, but a material of similar composition collects in swallow-holes and on the floors of caves, into which it is swept by the run-off from the rainfall.

In all probability the red clay-with-flints of some districts in south-eastern England is essentially a residual deposit derived from the Chalk, and related to terra rossa.

Laterite and bauxite

Laterites and bauxites with their thin cover of top soil are the extreme end product of the weathering cycle and are formed under oxidising conditions in areas of low or moderate topographic relief, but with a minimum of erosion. The climate is persistently warm or even subtropical with temperatures above 25 °C. Rainfall exceeds evaporation almost continuously and there is a progressive leaching of silica and alkali ions plus a destruction of humus by microflora. The presence of Al-rich parent rocks is preferable because it is the selective retention of aluminium which eventually creates the laterites and bauxites. The important aluminium hydroxides concentrated in these rocks are gibbsite, diaspore and boehmite and in the highest quality bauxites these minerals predominate. In laterites and lateritic and bauxitic clays there is a variable dilution by clay minerals, especially kaolinite, and iron oxides. The red and brown iron oxides in laterites sometimes form more than 25 per cent of the deposit and make the rock worth working as an iron ore. In south-east Asia the iron oxides frequently occur as pellets, spherules and nodular concretions embedded in a matrix mainly formed of kaolinite. The exposed surfaces present a slaggy or cavernous appearance and a brecciated structure is produced by the collapse and recementation of the superficial crusts.

Bauxite and laterite rocks are widely distributed throughout the humid tropics and subtropics. Well-known examples occur in Brazil, Jamaica, India and West Africa. The occurrence of similar soils or rocks has been observed in temperate latitudes, e.g. the northern shores of the Mediterranean, but laterite is never found where there is a prolonged cold winter. They result from the decomposition of a great variety of rocks of igneous, sedimentary and metamorphic origin. Most of the Indian laterite is found overlying the vast area of the Deccan basalts, but it is also found resting on gneiss and other rocks. In West Africa bauxite and laterite are found on schists, in East Africa on volcanic rocks and gneiss, and in South Africa on slate, sandstone and granite, whereas in Jamaica and Yugoslavia the parent rocks are limestones.

A feature of bauxite rocks is a pisolitic structure. The structure forms in the later stages of bauxitisation and seems to be created by the recrystallisation of feldspar derived from the parent rock.

Soils of subhumid and semi-arid regions

In temperate to cool regions, where the annual precipitation is between 250 and 600 millimetres, the rainfall is roughly balanced by evaporation. From the surface, effective leaching cannot take place, and there is a tendency for the mobile products of weathering, including some of the soluble salts, to remain within the soil. Calcium carbonate and gypsum are the commonest of the substances which accumulate in this way, but various other carbonates, sulphates, and other salts are also found in the soils. During the dry season water is drawn through the pore spaces to the surface of the soil, where it evaporates and deposits its dissolved salts. In the case of calcium bicarbonate, calcium sulphate, and colloids or soluble salts containing iron or silica, concretionary structures tend to be formed in the upper soil, and these substances thus accumulate in a fixed condition, without being washed down again during the moist season. In general, the clay fractions of semi-arid soils are base-saturated, usually with calcium, and consequently tend to have a mildly alkaline reaction.

The Chernozem of Russia owes its black colour to the presence of dark organic matter, which usually amounts to 8 or 10 per cent of the soil. The more soluble salts are leached out of typical black earths, but calcium carbonate and gypsum are incompletely removed, and tend to form concretionary masses at some level below the surface. Soils of this kind are characteristically developed on the extensive plains of the Russian steppes, where the parent material is principally loess, but they may also be formed from other rocks such as Jurassic clays, chalk, or granite. Similar black soils are found under comparable climatic conditions in Germany and the Great Plains of the United States of America.

The black cotton soil or regur of India is developed on the Deccan basalts under warmer conditions, and without the long and severe winters of the American and European plains. This kind of soil contains calcareous concretions known as kunkar, and generally resembles the black soils except in containing rather less organic matter. Similar black cotton soils are found also in Kenya and other parts of Africa.

The chestnut earths are so named from their dark-brown colour, which appears to be due to organic matter and not to ferric iron. They are found in parts of Russia and North America where the climate is rather drier than in the areas covered by black soils. Soils of this kind contain less organic matter than the black soils of similar latitudes, and calcium carbonate concretions accumulate nearer to the surface.

Soils of arid regions

Regions where the annual rainfall is less than 250 millimetres are typified by light-gray, brown and red alkaline soils. The paler colour of these soils is due to their poverty in organic matter, which often amounts to less than 1 per cent. Calcium carbonate and gypsum accumulate a short distance below the surface, the gypsum usually below the carbonate, and sodium salts are usually present in the soil. Where the water table is not far below ground level soluble salts are brought through capillary pores to the surface, where they become concentrated as the water evaporates. In this way efflorescences of soluble salts and nodular masses or encrustations of calcite and gypsum may be formed either at the surface or within the superficial layers.

In saline soils the preponderant salt is sodium chloride, which forms a white crystalline efflorescence at the surface during drought. These soils are most characteristically developed in arid or semi-arid regions such as

Turkestan, southern Russia, the Dead Sea basin, and parts of western North America. Sodium sulphate is usually present in saline soils and in some areas is the preponderant salt, as in the 'white alkali' soils of North America.

Alkaline soils contain sodium carbonate, and owing to the completely dispersed condition of their clay particles, they develop deep shrinkage cracks on drying; this imparts a characteristic prismatic structure to the soil. The black alkali soils of the arid regions of western North America are of this kind, and similar soils are found in Hungary and the Ukraine.

PALAEOSOLS

Several varieties of ancient sediment have qualities which are recognisably the product of soil-producing processes. Lateritic and bauxitic soils mantle weathered basic lavas of Carboniferous and Tertiary age in Scotland and Northern Ireland. Caliche layers are widespread in Old Red Sandstone and Lower Carboniferous molasse-type, red bed successions of the northern hemisphere. Podzolised profiles are present in Carboniferous alluvial–deltaic sequences and typical representatives are underclays, fireclays and ganisters (see p. 70), sometimes collectively referred to as seatearths.

Underclays and fireclays

Some authorities propose that the clays are direct weathering products of mudstone or shale rocks, the alteration being a combination of various subaerial leaching processes under humid climatic and acid conditions. Others suggest extensive weathering of mixed source rocks producing soil-colloids, colloidal silica and alumina gels which are then transported into the basin of accumulation and recrystallise into a clay rock complex.

Underclays and fireclays are either pale-gray or dark-gray with carbonaceous matter. They are frequently devoid of internal structure though slickensided surfaces may be common. Trace element content decreases towards the top probably due to leaching. In the Yorkshire and Durham Upper Carboniferous coalfields the general profile through the soils is:

Upper division   Generally carbonaceous with vitrain bands; some-
                  times bedded; plant remains and rootlets.
Middle division   Unbedded clay with rootlets and carbonaceous
                  films; slickensides common; iron carbonate nodules
                  towards the base.
Lower division   Transition zone of coarser material with occasional
                  rootlets, passing down into bedded shale.

Kaolinite laths, bundles and patches associated with quartz, illite and some chlorite are the main inorganic constituents and are intermingled with variable amounts of fungal, bacterial, algal and other organic matter. Representative analyses are given in Table 7.3.

Table 7.3 Chemical analyses of British Upper Carboniferous palaeosols.

	1	2	3
SiO_2	59·10	44·89	73·39
TiO_2	0·91	0·96	0·98
Al_2O_3	22·88	32·95	15·54
Fe_2O_3	2·69	2·31	0·45
FeO	—	—	1·49
MgO	1·80	0·94	0·73
CaO	0·26	0·25	0·28
MnO	0·01	0·02	0·02
Na_2O	0·71	0·03	0·25
K_2O	3·87	1·55	2·04
P_2O_5	0·03	tr	0·05
H_2O-	1·13	3·93	0·52
H_2O+	—	11·21	4·09
Organic C	2·53	0·58	0·19
CO_2	—	—	0·18

1 Thorncliffe, East Midlands coalfield.
2 Stannington, East Midlands coalfield.
3 Barsham, North Wales coalfield.

The kaolinite occurs as fine-grained matrix intermingled with iron carbonates and limonite, and as large vermicular grains which show evidence of replacing diaspore and other aluminium hydroxides. Some of the coarse kaolinite aggregates have a pisolitic or nodular structure probably inherited from early diaspore (hydrogen aluminium oxide).

The clays vary considerably in plasticity, some being very plastic and others hard and brittle. The very plastic varieties are gray, slickensided and may represent the lithification of relatively large clay mineral and aluminium hydroxide complexes. The non-plastic, flint clays, are smooth and white, and without slickensides; they probably formed from smaller sized colloidal complexes.

It should be realised that a fireclay, technically, should have refractory qualities, such as the ability to withstand great heat (up to 1600 °C) without fusion or disintegration. Many so-called fireclays do not have these properties.

The frequent occurrence of sideritic and sphaerosideritic nodules and aggregates implies a genetic connection with soil-producing processes (Fig. 7.6). It may be that leaching of iron in the soil by organic acids was the main mechanism of iron concentration at the local water table where stagnant reducing conditions prevailed.

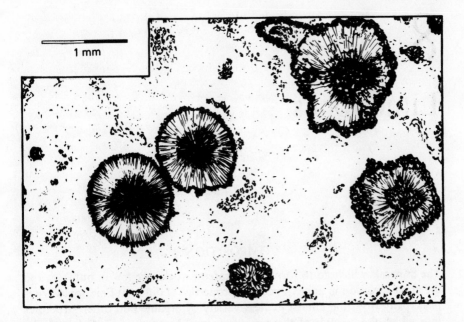

Figure 7.6 Fireclay, Carboniferous, central Scotland. An aggregate (clear) of aluminium hydroxides with minor amounts of clay minerals and fine silt. The sphaerosiderites are caused by secondary precipitation of iron. Polarised light.

8

Limestones

COMPOSITION

The minerals which go to form the calcareous rocks are few in number, and the great variation in appearance and properties of different limestones arises principally from the almost endless variety of organic and other structures into which the crystals of these minerals may be aggregated. **Calcite** is the stable form of calcium carbonate at ordinary temperatures, and may be regarded as the principal mineral of limestones. **Aragonite** is the form which calcium carbonate normally adopts when inorganically precipitated from sea-water. Conditions favouring its precipitation in preference to calcite are warm water, high alkalinity, supersaturation and an abundance of sulphate ions in solution. These conditions are usually met in the warm current areas of seas and oceans. Water rich in sulphate ions and trapped in sediments may help to preserve aragonite much longer than normal. Aragonite is unstable (or metastable) and under normal temperature and pressure conditions is liable to be converted into the more stable calcite. The older limestones normally contain no aragonite, and any shells which originally consisted of this mineral are found to be represented by open moulds or by coarsely crystalline (sparry) calcite. In most phyla, the shells consist either of calcite or of aragonite, and, except in certain mollusca, the two minerals are not found together in the same shell. Even in the complicated structures of molluscan shells, calcite and aragonite each build separate layers, and the crystals of the two minerals are never indiscriminately mixed.

Magnesium is found in many shells but it does not form separate crystals of magnesite, or of the double carbonate, dolomite. A review of a large number of analyses shows that magnesium only occurs to a very limited extent (less than 1 per cent) in skeletons consisting purely of aragonite, while it is much more compatible with the calcite lattice. In magnesian calcite shells the lowest percentages of magnesium are found in species inhabiting cold water. At water temperatures near to freezing point the $MgCO_3$ content in a range of common invertebrate skeletons varies between 0 and 6 per cent, whereas at water temperatures of 15 °C and higher the $MgCO_3$ content is often greater than 11 per cent and, occasionally, as much as 40 per cent. Calcite with less than 6 per cent $MgCO_3$ content is called **low-magnesian calcite**; with more than 11 per cent, **high-magnesian calcite**. Ferrous iron is also capable of ionic substitution in calcite lattices, giving rise to a group of minerals known as **ferroan calcites.**

Dolomite grains, $CaMg(CO_3)_2$, carry over 40 per cent $MgCO_3$ in contrast to the 40 per cent maximum in high-magnesian calcite grains. However, the total percentage of $MgCO_3$ in any given limestone need not necessarily reach a high figure before dolomite appears. Within Palaeozoic and Mesozoic rocks dolomite rhombs occur when the $MgCO_3$ reaches 2–3 per cent. It is probable that metastable high-magnesian calcites lose their magnesium with the passage of time and, by processes as yet obscure, form low percentages of dolomite.

There is complete isomorphism between dolomite and ferro-dolomite, $CaFe(CO_3)_2$, with ionic substitution of magnesium ions by ferrous ions in all proportions. The ionic radii of magnesium ions and ferrous ions are very similar and this eases substitution. Ferro-dolomite is rare in limestones, but there are frequent occurrences of stable carbonates intermediate in composition between dolomite and ferro-dolomite. These are generally called **ferroan dolomites** or **ankerites** and have theoretical formulae which are expressed either as $Ca_2MgFe(CO_3)_4$ or $Ca_3 (Mg_2Fe) (CO_3)_6$.

In general, ferroan calcites and ferroan dolomites are present as cements in limestones and tend to be formed late in any sequence of cementation. The minerals may be directly precipitated or may originate by ionic substitution within pre-existing non-ferrous cements.

Phosphatic minerals, when present in limestones, sometimes occur as an original part of organic skeletons, such as those of horny brachiopods or crustacea; they also take the form of early diagenetic and secondary concretions and replacements. The main phosphate is usually called **collophanite**, a mineral which falls into the hydroxyl-apatite–fluor-apatite series.

The accessory constituents of limestones include a great variety of substances apart from phosphates. Small grains of authigenic quartz are commonplace, authigenic albite less so. Microspherulitic pyrite is particularly prone to replace calcitic shell débris. Clay minerals of the kandite and illite groups are also intermittently distributed, some apparently being authigenic, others secondary emplacements within pores and cavities. Glauconite is present in many limestones, sometimes in considerable abundance, and occurs in both concretionary aggregates or replacements, and in the form of derived water-worn grains.

Non-calcareous organic remains are common, important amongst these being siliceous sponge spicules, and tests of radiolaria. Some of these organisms may be dissolved subsequently and reprecipitated as chert.

Pure limestones and dolomite rocks are white or very pale gray in colour, but even small traces of impurities will act as strong pigments, and may impart striking colours to the rocks. The most usual pigments are iron or manganese compounds, and finely divided carbonaceous matter (Table 8.1).

Table 8.1 Chemical analyses of limestones.

	1	2	3	4	5	6
CaO	54·82	54·84	53·81	54·70	49·49	51·03
MgO	0·30	0·26	0·56	0·60	1·07	1·20
CO_2	43·27	43·26	42·69	41·70	39·67	40·67
SiO_2	0·85	1·14	1·15	0·40	6·78	3·05
Al_2O_3	0·18	0·41	0·45	0·52	0·75	0·35
Fe_2O_3	0·13	—	—	0·08	0·47	0·01
MnO	0·04	—	—	—	0·09	0·01

1 'White' chalks, England (calcilutite), 9 analyses.
2 Carboniferous Limestone, North Wales (calcarenite).
3 Lithographic Limestone, Solenhofen, Bavaria (calcilutite).
4 Salem Limestone, Carboniferous, Indiana (oolitic calcarenite).
5 Cornstone (caliche), Old Red Sandstone, Scotland (sandy calcisiltite).
6 Shell-sand, Recent, Scotland (uncemented calcarenite).

CLASSIFICATION

Some classifications are based on characteristics determinable in the field only; others are based on binocular inspection or thin-section studies. The terminology appropriate for unlithified modern deposits is difficult to apply to fossil deposits and vice versa and the progressive diagenetic changes, to which carbonate rocks are particularly susceptible, may be such that genetic classification becomes hazardous. No classifications are completely successful and may be criticised to a greater or less degree. On the other hand, the balanced combination of descriptive and genetic parameters within certain modern classifications make them suitable for universal use.

Grain size is a useful basis for subdividing clastic limestones, as in Figure 8.1. The system adopted is based on the Wentworth scale for fragmental deposits (Table 2.1) with suitable modifications. A disadvantage of the scheme is that it is uninformative about the nature of the constituents, although appropriate descriptive names can be added, such as oolitic calcarenite and algal calcisiltite.

The classifications of Folk and Dunham put more emphasis on constituents and texture than on grain size. They are more explicitly descriptive than some schemes and have the added advantage of indicating textural elements from which depositional and diagenetic processes can be deduced. A point worth emphasising with these classifications is that the name adopted applies to

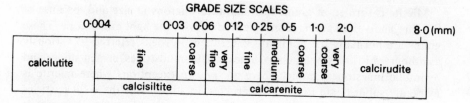

Figure 8.1 Clastic limestone classification.

the sediment as seen now and this appearance commonly post-dates a whole host of early and late diagenetic changes, which may have considerably modified the original deposit.

Folk's groups are:

I   Allochemical   — sparry calcite (sparite) cement
II Allochemical   — microcrystalline aragonite and calcite (micrite) matrix
III Orthochemical — micrite main constituent
IV Autochthonus reef rocks
V   Replacement dolomites

The basis of the grouping is that most unaltered and partially dolomitised limestones consist of three identifiable components – allochems, micrite and sparite.

Allochem is a collective word for all varieties of discrete carbonate aggregate most of which have undergone transport at some stage in their history. The most important allochems are **intraclasts, pellets** (peloids), **ooliths** or **ooids** and **shells** (skeletal débris). Intraclasts are limestone or dolomite fragments of all sizes formed by pene-contemporaneous erosion of either the adjacent sea floor or exposed carbonate mud flats but do not include fragments derived from older limestone outcrops. They range from sand-size grains to pebbles and boulders. Some fragments may be derived from contemporary algal growth, become rounded then subject to renewed peripheral oolitoid algal growth. They are called **oncolites**.

Other small rounded grains, less than 0·15 mm, in size, may consist of homogeneously fine-grained limestone and are difficult to differentiate from pellets, which have an entirely different origin. Pellets are ovoid to spherical grains, usually between 0·04 and 0·08 mm in size, which have no ordered internal structure unlike ooliths and most intraclasts. They are commonly of faecal or algal origin but, if the origin is indeterminate, the non-genetic term peloid is preferable. Because of identification difficulties, some classifications based on Folk's scheme do not distinguish between pellets and intraclasts.

Micrite is formed of grains less than four microns in size and, because of this fact, individual grains cannot be studied with the light microscope. These grains are normally the dominant constituent in rocks referred to variously as lithographic limestones, chinastones, calcite mudstones, cementstones and hydraulic limestones, though there are known exceptions where micrite is markedly subordinate to silt-size particles. In some cases these silt particles are almost certainly detrital but in others, especially where the grain size is 5–10 microns, it is possible that they have formed by small-scale recrystallisation of micrite mud. The latter process produces grains sometimes referred to as **microspar**. The Devonian Griotte and Cephalopodenkalk pelagic limestones of southern France and Germany have been intensively converted into microsparite rocks. To complicate matters even further there is a diagenetic process affecting lime deposits known as 'grain diminution' in which carbonate particles of all sizes are altered early diagenetically into new or neomorphic micrite. Because of these complications it is often difficult in practice to distinguish between, not only the origins of micrite grains but also the sizes of individual small grains.

Sparite consists of clean, coarse-grained calcite grains more than ten microns in size. The grains infill pre-existing cavities or pores within a limestone and if these should be large the grains may reach a size well over 1 mm. They frequently infill shell cavities; sometimes the whole cavity, sometimes only the upper part, the lower being occupied by fine-grained carbonate detritus. Sparry calcite is also common in limestones formed dominantly of ooliths or well sorted, well rounded shell débris. There is a tendency for the individual calcite grains to increase in size and decrease in number away from the allochems they are cementing towards the centre of the original pores and cavities.

Sparry calcite has to be carefully distinguished from recrystallised calcite, the grains of which may also be larger than ten microns. Recrystallised calcite is a secondary replacement of earlier carbonates and consequently tends to transect the boundaries of pre-existing textures and structures. In the absence of such obvious features identification becomes very difficult and dependent on a range of criteria, such as wavy grain boundaries, floating relics and variably patchy grain size, not all of which may be present or detectible. Where replacement is extensive only 'ghost' traces of the original textures may remain as evidence of alteration. The grain size within recrystallised calcite patches is very variable possibly as a consequence of pressure solution effects. Differential pressures along the boundaries of original calcite grains give rise to solution at points of strain and reprecipitation at points of least pressure. In this way the original grains change shape and selectively enlarge until a patchy secondary mosaic of irregular shaped grains is created. One product of this type of alteration is **pseudo-breccia**. Lower Carboniferous pseudo-breccias from northern England and Wales exhibit 'fragments' which are dark gray in colour and set in a pale gray micrite matrix. Thin-section inspection shows that the matrix contains fossils whereas the 'fragments' consist predominantly of irregularly shaped patches of coarse calcite. These patches, in fact, are simply areas of more advanced recrystallisation. Occasionally, crinoid plates 'floating' within these calcite patches have enlarged by marginal growth at the expense of the matrix. Under these circumstances, the growth of new calcite on the crinoid margins may be in structural and optical continuity – an effect known as syntaxial or rim growth. The cementation ensuing from syntaxial growth is sometimes called 'rim cementation'. Multiple consecutive phases of rim cementation have been observed affecting shell fragments in certain limestones.

Figure 8.2 illustrates how the three main components of limestones plus the variation in the nature of the allochems is used for subdivision within each of Folk's groups. The rocks are named by combining, in abbreviated form, their major petrographic attributes. The first part of the hybrid name refers to the allochem component and the second part to the cementing or

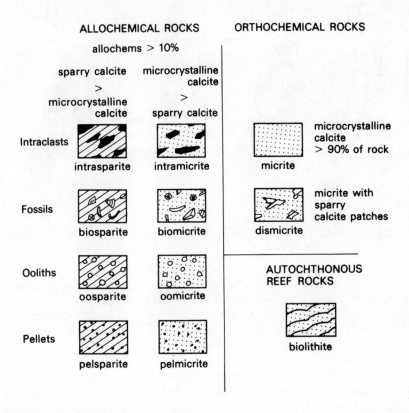

Figure 8.2 Folk's classification of limestones. Application of the terminology to the main varieties of limestone.

matrix materials. For example, intrasparite consists of intraclasts cemented by sparry calcite. Biomicrudite is a further style of modification in which the grain size of the dominant allochem is taken into account. In this particular rock the average size of the shell fragments set in a micrite matrix is greater than 2 mm.

The use of Folk's method of classification has been demonstrated on the modern sediments which mantle the Andros Platform in the Bahamas (Fig. 8.3). It is interesting to compare the areal distribution of his limestone types with that of a different, older style of carbonate classification. The main discrepancies between the two systems arise in the mud areas immediately to the west of Andros Island and are due, in part, to inherent differences in the methods of classification.

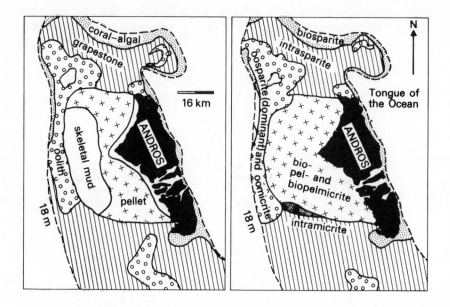

Figure 8.3 Sediments around Andros Island, the Bahamas. A comparison of conventional and Folk's terminologies when applied to modern carbonate sediment on the bank. The eventual nature of the cement is assumed in the sparite zones. Also note that grapestones are not now regarded as being a variety of intraclast, so the intrasparite zone is a misnomer; the older name 'grapestone facies' remains more suitable. Maps after Imbrie and Purdy (1962).

Dunham's classification is essentially textural and is most valuable when used in a purely descriptive way for lithified rocks (Table 8.2 and Fig. 8.4). Textural maturity is implied in that the least mature varieties are richer in mud matrix than the more mature grainstone varieties. However, depositional deductions based on these textures alone need great care.

Table 8.2 Classification of carbonate rocks according to depositional texture.

Depositional texture recognisable					Not recognisable
Contains mud		Grain-supported	Lacks mud and is grain-supported	Original components bound together	CRYSTALLINE CARBONATE
Mud-supported					
less than 10% grains	more than 10% grains				
MUDSTONE	WACKESTONE	PACKSTONE	GRAINSTONE	BOUNDSTONE	

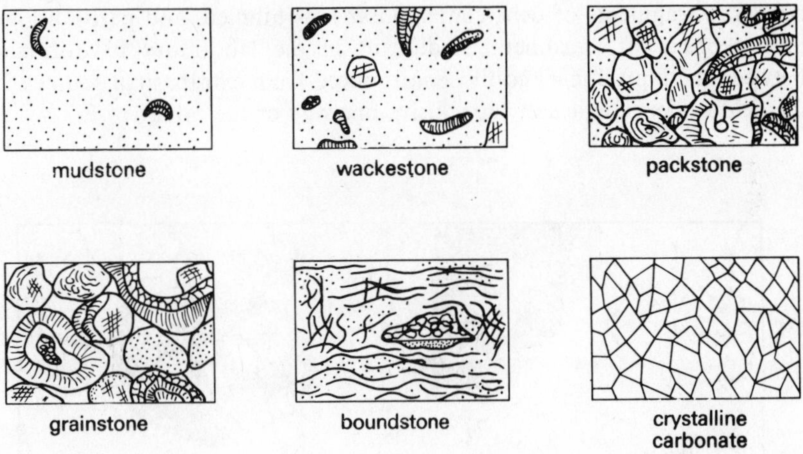

Figure 8.4 Dunham's classification of limestones. The fine stipple represents mud matrix.

The term mudstone is not quite synonymous with micrite or calcilutite as it includes all grains up to twenty microns in size. Wackestone consists of more than 10 per cent of grains greater than twenty microns in size 'floating' in a matrix of mud, and packstones are composed of grains in close contact with each other with interstitial mud cement. In analysing the grain size distribution of modern sediments it is found that the lutite and silt-grade materials are often so inextricably mixed that division into those above and below twenty microns in size is very difficult. Lutite and silt may then be grouped together as 'mud' and the term 'grain' reserved for sand-sized and larger particles (> 60 microns). Grainstones are mud-free carbonate rocks indicative of relatively strong bottom currents. To make the classification comprehensive the term boundstone is used for rocks formed mainly of binding, net-like organisms, such as reefs, and crystalline carbonates for those in which nearly all original texture is lacking, such as dolomites.

ALLOCHEMICAL LIMESTONES

Within this largest of all limestone groups occur a range of rocks which are predominantly clastic in origin. Only a selection can be discussed here.

Intraclast-bearing varieties

These comprise endogenetic rock fragments up to boulder size, which are set in a primary micritic matrix and/or secondary sparite cement (Fig. 8.5). The style and degree of cementation by sparite only become apparent on lithification. True intraclasts are intrabasinal products and are essentially reworked fragments of pene-contemporaneous lithified, and partly lithified, lime deposits. It is sometimes difficult in the laboratory to distinguish between these products and those introduced from extrabasinal sources, but the field evidence is usually conclusive one way or the other.

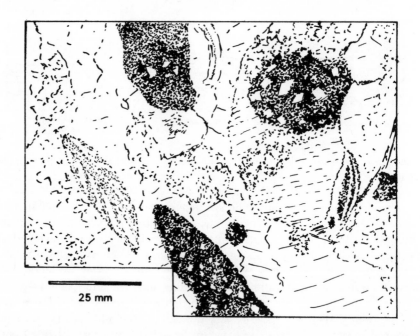

25 mm

Figure 8.5 Intraclastic limestone, Palaeozoic, Wyoming. Intraclasts of dense micritic limestone and shell fragments set in a sparite cement. Small dolomite rhombs are scattered through the rock. Polarised light. After Dixon and Reeves (1965).

The roundness of intraclasts varies considerably, depending on the amount of transport. Some supratidal and lacustrine mud-flake deposits contain highly angular fragments. So do some of the extensive submarine talus deposits associated with modern and ancient reefs. In contrast, reworking in protected, shallow water environments can give good rounding. Aeolian processes, effective in moving sand-sized particles from the littoral into the beach and dune zones, as in the Persian Gulf, also produce good rounding.

The quality of sorting in intraclastic limestones is as variable as the particle roundness factor. In shelf areas with restricted current activity or in deeper waters where turbidity current emplacement has occurred, the sorting may be poor and the quantity of micritic matrix relatively high. The textural characteristics are then of the wackestone–packstone type. Supratidal carbonates commonly have a wackestone texture. The slopes of deep basins and troughs (e.g. Tongue of the Ocean, Fig. 8.3), bounding shallow water areas of lime accumulation, are the loci for graded bed deposition, in which the basal packstone layers of each graded unit are sometimes pebbly, with genuine intraclasts and exotic clasts. Turbidite beds of this kind have been called 'allodapic limestones'.

The cleaner washed, well sorted intraclastic limestones generally typify shallow water, strong current and wave-affected environments. In these situations any pre-existing micrite is winnowed-out and sparry calcite precipitated in suitable interstices, as soon as the fragments are mechanically stabilised. The texture usually falls into the grainstone category. Ripple marks, cross-bedding and scour structures characterise both modern and ancient beds.

Grapestones are lumpy clusters of strongly micritised and algal-bored carbonate sand grains, cemented by micritic aragonite. They used to be regarded as a variety of intraclast. It is now known that they form, probably as a consequence of in-place micrite precipitation around grains immobilised in algal-rich mats. These subtidal mats commonly cover large areas of the sea bottom, as in the northern and southern parts of the Andros Bank, and are pale brown-green layers, a few millimetres thick and rich in microorganisms. Filamentous algae and diatoms are dominant and trap and bind the clastic particles into a gel-like coherence. The layers are moderately resistant to erosion and stabilise the sediment underneath to a much greater degree than would be possible otherwise.

Biosparites

Among the well bedded marine limestones, the most widespread type consists of broken or disintegrated fragments of large calcareous skeletons, mixed with the complete shells of smaller organisms, and cemented together by comparatively clear sparite. Many grainstones are of this type.

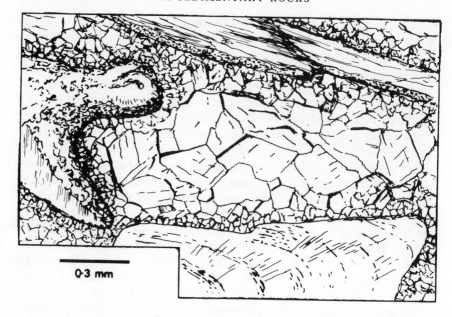

Figure 8.6 Biosparite, Upper Lias, south England. The shell fragments have, in places, a micritised veneer. A thin, irregular zone of neomorphic microspar surrounds the fragments, and this zone passes abruptly into a central pocket of clear sparry calcite. Polarised light.

Frequently, in the well sorted varieties, it will be noticed that all the recognisable fragmental remains are those of calcitic organisms. Possibly the carbonate of any original aragonitic organisms has been dissolved, redistributed in solution and reprecipitated in pores and cavities as early diagenetic sparite cement. This mechanism for cement production, however, is likely to be too simple, as it does not take into account the probability that some of the coarse calcite is not cement technically, but is neomorphic sparite, formed by in-place replacement of original carbonate. Certain thin Liassic limestone beds in Britain and parts of the thicker, more massively bedded Lincolnshire Limestone (Middle Jurassic) of the Midlands of England have sparry calcite of mixed origin (Figs 8.6 and 8.7). In general, the greater the lithification and age of a biosparite the more it should be suspected that mixed origin sparite is the rule rather than the exception.

Although shell débris (and other allochems) in biosparites are frequently recognisable by their morphology or specific internal structure, it has to be appreciated that the fragments may have been modified on the sea bottom prior to final incorporation into the rock. This modification is commonly initiated by the activities of fungi and endolithic non-calcareous algae, which bore into the surface of the fragments and create a dense network of tubes,

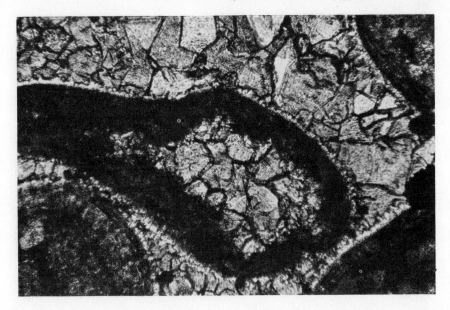

Figure 8.7 Biosparite, Middle Jurassic, central England. Clear sparry calcite cementing and infilling the cavities of micritised shell fragments. The shells have an outer veneer of microspar in places. The variation in grain size of the coarse sparite cement suggests that a certain amount might be neomorphic. Polarised light, × 30.

each tube being up to about six microns in diameter. Within the tubes, possibly aided by bacterial activities, conditions are established suitable for the precipitation of micrite. The process of micritisation has the tendency, therefore, to destroy original organic internal structures. So, some rolled micritised shell fragments may begin to closely resemble faecal pellets in shape and composition.

Crinoidal biosparites are rocks in which the ossicles of crinoid stems are conspicuous ingredients. Each ossicle consists of calcite which behaves as part of a single crystal, and breaks along cleavage planes. Syntaxial rim cementation is common. Consequently, a fractured surface of the rock may give a deceptive appearance of a coarsely recrystallised limestone, owing to the abundant large and conspicuous cleavage surfaces. This effect is particularly striking in some of the crinoidal rocks associated with the Wenlock Limestone (Silurian) of the Welsh Borderland and the Lower Carboniferous successions of Wales and England (Fig. 8.8).

Modern shell accumulations which might lithify into biosparites are found in various intertidal and subtidal situations. In the shallow waters on the south-western flanks of the Persian Gulf there are extensive areas of clean washed mollusc, foraminiferal, algal and coral sands, where conditions

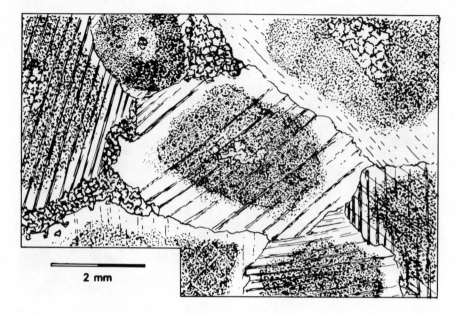

2 mm

Figure 8.8 Crinoidal limestone, Carboniferous, Forest of Dean, England. Dusty looking crinoid ossicles enlarged by syntaxial (rim) cement. The outgrowths are in optical continuity with the calcite of the ossicles. There are small interstitial pockets of microspar cement. Polarised light.

appear appropriate for the precipitation of interstitial calcite. The rate of cementation depends on many obvious factors, such as the geochemistry, temperature and movement of the pore waters, also less obvious factors such as shell mobility. If the deposits stabilise for some reason then cementation is likely to occur much faster.

Biomicrites

Certain limey deposits consist of organic shelly detritus embedded in a micrite matrix; they are biomicrites or shelly wackestones. Deposits of this kind form in quiet water, where fine-grained carbonate mud can settle, and organic remains are entombed in an unbroken condition, irrespective of size or fragility. The waters can be relatively deep, for example below forty metres in the axial parts of the Persian Gulf, or very shallow and intertidal, as in the coastal zone of the same region. Much of the shelly part of the Wenlock Limestone (Silurian) is shallow water biomicrite and probably accumulated in sheltered pools among the reefs of stromatoporoids and corals.

The fossiliferous bands of the Cambrian Durness Limestone in north-west

Scotland consist principally of shells of molluscs embedded in a fine-grained matrix. A parallel lithology is also found in many of the ammonite-bearing limestones of Jurassic age in southern Europe. The chambered shells are frequently preserved as internal casts formed of micrite or sparite, and the limestone beds themselves often have an indistinct nodular structure, suggestive of early irregular diagenesis.

Algal limestones

Calcareous algae secrete a hard limey skeleton. Two families, the Corallinaceae (red algae) of Carboniferous to Recent age and the Codiaceae (green algae) of Cambrian to Recent age, are much more important than any others at the present day, and are confined to marine or brackish waters. The Corallinaceae require a hard foundation upon which to grow, and are principally found off rocky coasts or in coral reefs. The skeleton consists of closely packed, parallel calcified tubes, commonly formed of high-magnesian

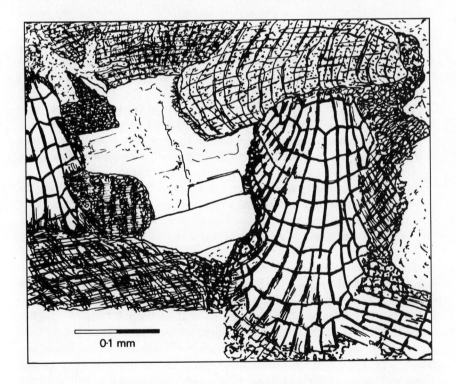

0·1 mm

Figure 8.9   Lithothamnium limestone, Eocene, Bavaria. Two species of the red alga *Lithothamnium* are present and set in a sparite and micritic matrix. Polarised light.

calcite in modern genera. These tubes are usually only discernible through a microscope; longitudinal sections show well marked concentric growth surfaces. The family is important in modern seas and in Tertiary deposits, some of the best represented genera being *Lithothamnium*, *Archaeolitho-thamnium*, *Amphiroa* and *Lithophyllum* (Fig. 8.9). The remains of these plants are sometimes found as rolled fragments and sometimes as encrusting or frondose masses preserved in the position of growth.

In calcareous rocks older than the Cretaceous, the Corallinaceae are less prominent than the closely related red algae Solenoporaceae. *Soleno-pora* is an important rock-former at certain levels in Carboniferous lime-stones of northern England, and *Solenoporella*, although a locally restricted fossil in Britain, is closely associated with Jurassic coral reefs in much the same way as *Lithophyllum* is associated with modern reefs.

The Codiaceae, such as *Halimeda*, *Penicillus* and *Udotea*, are important contributors to the deposits of modern tropical seas, but their remains are not found in any appreciable quantity in older limestones. These forms, unlike the Corallinaceae are capable of colonising unconsolidated muddy bottoms, and are consequently able to live in environments which would be impossible to the Corallinaceae. They are firmly established on the Bahamas Banks where *Halimeda*, on disintegration, contributes sand-sized, often micritised, grains to the sediment. In contrast, *Penicillus* and *Udotea*, which appear to be constructed of more weakly aggregated aragonite needles than *Halimeda*, tend to break down into fine grained mud.

From the Cambrian to the Miocene the principal calcareous green algae belong to another family, the Dasycladaceae, which are poorly represented in British rocks, but make up a large part of the Alpine Mesozoic limestones in south-east Europe and Arabia. Dasyclads generally thrive under conditions of relatively strong current action, with little accumulation of micrite.

Some algae, especially filamentous cyanophytes, such as *Schizothrix*, do not necessarily secrete a calcareous skeleton, but in certain circumstances become encrusted in calcium carbonate. If these forms grow under water they will only precipitate calcium carbonate if the water in contact with them is saturated with this substance. The algal nodules known as 'water biscuits' or 'lake balls' and found in modern fresh-water lakes in many parts of the world, are formed in this way. Similar structures also occur in the deposits of Tertiary fresh-water lakes and lagoons. The modern fresh- and marine-water genus *Scytonema* sometimes forms cushion-shaped tufts of roughly parallel tubes, which may become calcified, almost indistinguishable in thin section from colonies of the Palaeozoic genera *Ortonella* and *Mitcheldeania*. It is also probable that *Girvanella* and *Sphaerocodium* are remains of species which only became calcified when growing in water saturated with calcium carbonate.

Cyanophyte algae are often among the earliest colonisers of sediment newly deposited on intertidal mud flats. It is well known that such algae exert a strong stabilising effect upon the sediment they colonise, largely because they bind it together with their growing filaments, and ultimately cover it with a mass of felted tubes. On intertidal flats, where there is an abundance of drifting mud and silt the algae tend to collect this sediment and progressively create a discontinuous, irregularly laminated and mat-like structure, known as a **stromatolite**. These structures cover extensive areas of the high flats in the Bahamas, south Florida and Persian Gulf. In the more highly saline parts of some embayments, such as Shark Bay in Western Australia, they extend into the lower flat zone. The individual laminae are rarely more than a few centimetres thick and it would be unusual for a whole structure to have a thickness greater than one metre. Internally, the layering is often complex with mound-like, 'algal head' morphology.

Structures comparable with modern littoral stromatolites are well developed in ancient shallow water limestone successions. Ovoidal, dome-like structures as much as fifteen metres across and one metre high are present in the Middle Magnesian Limestone (Permian) reef facies of north-east England. The lamina frequency varies between 36 and 48 per centimetre thickness. The largest and most complex growth forms are associated with the reef crests and upper reef slopes, whereas the smaller, less complex varieties appear analogous to those forming on modern reef flats and in adjacent lagoons.

Some late Precambrian and Palaeozoic limestones show traces of cellular or filamentous structure. These organic characteristics, seen particularly in the growth forms Collenia and Cryptozoon, suggest the activities of sediment-binding algae, similar to the modern cyanophytes. These algal structures are of special interest since they can build up reef-like masses. Particularly fine Precambrian developments are to be found in the Algonkinian limestones of Montana and in the Transvaal Dolomite of South Africa.

The minute surface-floating algae grouped together as the Coccolithaceae (Jurassic to Recent) each consist of a single, approximately spherical cell encased in a sheath of low-magnesian calcite plates. The plates vary in size between 2 and 20 microns. The algae have a wide distribution in the sediments of temperate and tropical oceans, and are specially abundant in the globigerina ooze of the North Atlantic. The spinose rhabdoliths (another variety of coccolith) are, on the other hand, confined to warmer waters, such as the Mediterranean and the tropical parts of the open oceans.

Chalk

This is a very special variety of biomicrite, characteristic of the Upper Cretaceous of western Europe and parts of North America. It is, as a rule, a remarkably pure rock containing more than 90 per cent calcium carbonate (Table 8.1). Dolomite is rare. The white colour and high reflectivity to light indicate its purity and fine-grained constitution. Some 75–90 per cent of the carbonate is in the form of organic particles less than four microns in size.

The main impurities are clay minerals of which montmorillonite and illite are most common. Glauconite is widespread, but in minute quantities. It is not known whether all, or just some, of these minerals are authigenic. This contrasts with pyrite granules and nodules, and small alkali feldspars, which are undoubtedly authigenic. Authigenic collophanite is sometimes concentrated as impregnations, grains and nodules at levels which are believed to indicate very reduced or zero rates of deposition.

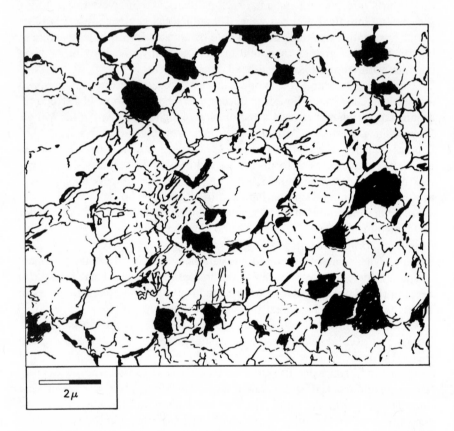

2μ

Figures 8.10 Chalk, Cretaceous, Northern Ireland. A ghostly trace of a coccolith in the chalk, as seen using an electron microscope.

The bulk of chalks are made up of the remains of planktonic marine algae, the Coccolithaceae, which can form up to 80 per cent of certain beds (Fig. 8.10). There are many varieties. Preservation of the oval coccolith discs is occasionally good, but physical degradation, compaction and other diagenetic effects generally break them down into smaller fragments and discrete crystal laths.

The rest of the calcareous organic débris consists of whole or worn fragments of a wide range of invertebrates. Molluscs, echinoderms, bryozoa and foraminifera are the most common of these fossils and some larger specimens may be completely silicified.

The texture of the chalk varies considerably from bed to bed. A high proportion of *Inoceramus* prisms gives rise to a perceptibly gritty-textured rock, whilst a reduced quantity of molluscan débris, and a correspondingly large bulk of coccoliths, results in the finest grained varieties.

In addition to foraminifera, the chalks contain spinose spherical bodies, up to 500 microns in diameter, which superficially resemble large isolated chambers of globigerina in appearance. The central area is usually occupied with sparite cement. A foraminiferal origin is nevertheless improbable, since they possess no aperture and appear to be imperforate. These **calcispheres** are especially abundant in certain hard or nodular bands. They are most probably the remains of some planktonic organism, possibly akin to modern dasycladacean algae (Fig. 8.11).

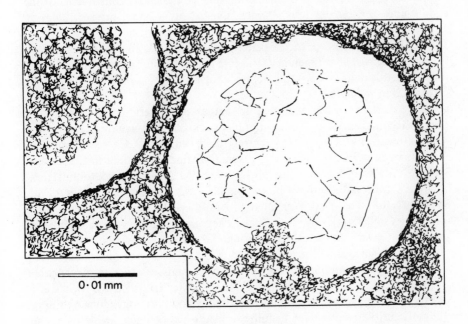

0·01 mm

Figure 8.11 Calcisphere in the Chalk, Cretaceous, south-east England. These organic structures are readily visible using an ordinary microscope. They show commonly a narrow, clear outer zone with a core occupied either by micrite (top left) or microspar (centre right). The periphery of the spheres is often irregular when inspected closely and appears to be 'corroded' by the fine chalk matrix. Polarised light.

Remains of aragonitic organisms are absent from the soft chalk, but in certain hard bands, such as the Chalk Rock, which appear to have suffered lithification at an early stage, aragonitic shells such as those of ammonites and gastropods are represented by hollow moulds.

It has been suggested that, unlike other limestones, the chalk was originally laid down as a low-magnesian calcite-rich not aragonite-rich mud. Hence, no recrystallisation of aragonite to calcite has subsequently occurred, and so the rock has retained a high porosity (36–65 per cent).

In Scotland, Antrim and north-eastern England the chalk is moderately well cemented by secondary calcite infilling original pores and forms a comparatively hard limestone, whereas in southern England it is much softer and more friable. The density of the Antrim chalk varies between 2·60 and 2·64 and that of the soft English chalks between 1·70 and 1·95 because of this difference in cementation. The reasons for the peculiar geographical distribution of the hard and soft chalks is obscure but it has been suggested that the hard layers originally had a greater proportion of micrite deposited between the larger organic particles. This micrite may have recrystallised into syntaxial grain growths around the margins of the larger constituents of the chalk and so created a tougher rock. On the other hand, pressure solution between the sediment constituents may have provided syntaxial cement without fine micritic particles being involved at all. In the Antrim chalk there is evidence of sedimentary structure deformation, fracture of fossils and interpenetration of shell fragments, which would suggest hardening by overburden weighting.

On a different much smaller scale, hardening of certain beds could have been accomplished by contemporary submarine cementation processes. **Hardgrounds** are layers, a few metres thick, where the degree of aragonite and magnesian calcite cementation, some of which is neomorphic, decreases downwards from the depositional interface. At present, hardgrounds some 5–10 cm thick are forming in the shallow water carbonate zones of the Persian Gulf. The chalk equivalents in southern England are given names, such as Chalk Rock and Top Rock. They are commonly bioturbated, and pebbles, organic encrustations, phosphatisation and glauconitisation typify their upper surfaces. Erosion, reduced rates of deposition and a shallowing of the sea over a period lasting several thousands of years seem probable.

The manner of deposition and depth of water in which the varieties of chalk were deposited are difficult to estimate from lithology alone. Although lamination, flat and slump bedding, and erosive contacts are known, they may be rendered almost unrecognisable at levels of intense bioturbation. Cross-bedded carbonate banks in the Normandy chalks, with amplitudes as much as fifty metres and up to 1·5 kilometres long, are indicative of considerable bottom currents.

Faunal and floral considerations put a minimum depth limit for the finest chalks at about 100 metres. The bulk of the coccolith fragments show no 'corrosive etching', which suggests deposition at depths no greater than 600 metres. The evidence of the molluscan fauna of the hardgrounds indicates depths less than 200 metres.

A curious feature of the Cretaceous chalk beds of western Europe is the development of a distinctive brick-red staining at certain marly, nodular and bioturbated levels. In eastern England the zones of reddening are usually about one metre thick, though they can reach six metres locally. There is evidence of the staining transecting primary bedding features but the distribution, in general, is concordant with bedding suggesting a contemporary or pene-contemporaneous origin. The colour is caused by granular haematite, up to 40 per cent in quantity, which is distributed unevenly throughout the zones. There is a possibility that some of the haematite is secondary after pyrite and glauconite, but how much remains conjectural.

The conditions giving rise to the reddening are also problematical. Suggestions have been made that the cause was the introduction of clastic lateritic mud into the Chalk Sea.

Globigerina ooze

This is one of the most widely distributed of all deposits, since it is estimated to cover an area of nearly 125 000 000 square kilometres. It attains its maximum development in the Atlantic, where equatorial sedimentation rates are

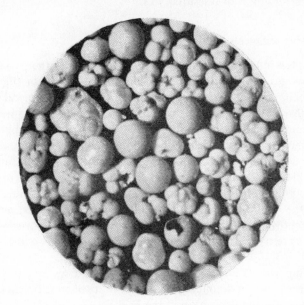

Figure 8.12 Globigerina ooze, Recent, Arabian Sea. Sample obtained from a depth of 3054 metres. The organic remains were set in a brownish-gray matrix. × 15.

of the order of $3 \cdot 5$ cm per 1000 years, and also covers great areas in the Pacific and Indian Oceans; it is known to extend as far south as latitude 60°, and as far north as the Arctic Circle. The globigerina ooze occurs at all the medium depths of the ocean removed from continents and islands and is especially developed in those regions where the surface of the sea is occupied by warm currents and is organically very productive.

At depths greater than about 4500 metres, the concentration of calcareous organisms in the oozes falls rapidly to very low values. This is because the waters are markedly undersaturated with respect to calcium carbonate, so that solution of the tests of organisms occurs relatively quickly.

Globigerina ooze is a cream-coloured, pink, or pale gray substance, muddy when wet, and powdery when dry; it consists for the most part of the tests of foraminifera, of which various species of *Globigerina* are the most abundant (Fig. 8.12). Many other organisms also occur, especially pelagic molluscs – pteropods and heteropods – and calcareous algae. Siliceous organic remains are sometimes present in considerable quantity and consist chiefly of diatoms, radiolaria, silico-flagellates, and the spicules of sponges. Some samples dredged up contain relatively high proportions of celestobaryte grains – a solid solution of celestite ($SrSO_4$) and anglesite ($PbSO_4$) in barytes ($BaSO_4$). The probability is that the calcareous organisms extract these materials from the near-surface waters in the construction of their skeletons. On death, skeletal disintegration at depth leads to the accumulation of the discrete mineral grains of celestobarytes.

Oolith-bearing sediments

These include a number of carbonate rocks agreeing in the possession of a characteristic texture somewhat resembling the roe of a fish (Fig. 8.13). A normal oolitic deposit is made up of an aggregate of spherical allochems called **ooliths** or **ooids**, usually one millimetre or less in diameter. When examined under the microscope, the grains are found to be built up of two or more concentric layers, usually around a nucleus, which may consist of a shell fragment or a grain of authigenically modified clastic quartz. The material which makes up each of these concentric layers consists, in most modern deposits, of aragonite crystals laid roughly tangentially to the surface of the oolith. In many of the older oolitic limestones, on the other hand, the ooliths are found to be made up of little radiating prisms of calcite, created by internal recrystallisation, set at right angles to the concentric rings.

All the important developments of oolitic limestones among Pleistocene and older rocks are undoubtedly marine or salt-water sediments. Modern occurrences are in the Persian Gulf, the Red Sea, on the shallow submarine platforms of Florida and the Bahamas, and on the shores of the salt lakes of the western United States, such as the Great Salt Lake of Utah.

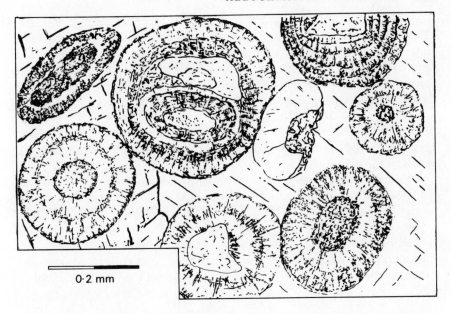

0·2 mm

Figure 8.13 Oolitic limestone, Middle Jurassic, central England. An oosparite according to Folk. The characteristic radial and concentric structure can be seen in most of the ooliths. Some of the ooliths have a single shell fragment nucleus, whereas others have a compound nucleus (top centre). In some instances, such as at the top left, there seem to have been phases of micritisation of the ooliths between phases of outgrowth. Polarised light.

In Florida and the Bahamas (Fig. 8.3) ooliths are growing where water saturated with calcium carbonate is drifted backwards and forwards over a shallow sea floor in the track of tidal currents. The depth of water is never very great, usually not more than six metres, and the sea floor is so scoured by currents that all traces of micrite are commonly winnowed away. The grains which remain are kept in motion, and, acting as nuclei, become encased in concentric layers of carbonate precipitate.

The oolitic sands of the Gulf of Suez (Red Sea) appear to be forming in the littoral belt (between tide-marks). They form wide stretches of yellow, ripple-marked sand, exposed at low tide. When thus exposed to the tropical sun, part of the sand dries, and tends to be blown inland as small dunes, which advance towards the desert. Around the Persian Gulf, blown oolitic sands are recognisable some forty kilometres away from the shoreline. The name **aeolianites** is applied to these internally cross-bedded deposits. The nuclei most commonly consist of grains of quartz, feldspar, or of foraminiferal shells.

Oolitic sands with a more than 50 per cent oolith content are forming at the present day in the highly saline Laguna Madre of the Texas Gulf Coast. The shoreline areas of accumulation are constantly wave-washed, have low rates of terrigenous deposition and are subject to high summer temperatures and evaporation rates. Hence, the waters are carbonate saturated and precipitation of aragonite occurs freely, commonly building up eight to ten layers around detrital nuclei. Within the individual layers of the ooliths aragonite fibres are arranged radially, concentrically or are completely unorientated. The reasons for this variability in fabric, which is recognisable in ancient ooliths despite conversion of the aragonite into calcite, remain a mystery.

Modern marine oolitic deposits usually accumulate in very shallow agitated waters and, because of this, are most likely to be cemented by sparry carbonate once the ooliths come to rest for an adequate period. Under some circumstances, bottom currents transport ooliths into less disturbed waters and they become incorporated into micritic mud. Alternatively, due to some change in current paths, mud may be introduced into the interstices between the ooliths. In both instances the growth of individual ooliths

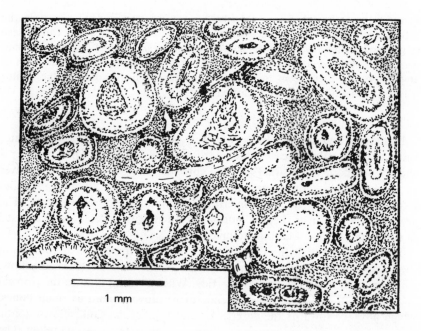

1 mm

Figure 8.14 Oomicrite, Lower Carboniferous, Morocco. Ooliths and one-layer superficial ooliths set in a micrite matrix. As the ooliths show little evidence of abrasive wear, it is possible that the micritic mud filtered down between the ooliths during a subsequent, much quieter phase of carbonate deposition on the sea bottom. Thus, the matrix is younger than the ooliths. Polarised light.

virtually ceases and the sediment is prospectively an oomicrite or oolitic packstone (Fig. 8.14).

There is considerable evidence that ancient oolitic limestones were mostly formed in shallow, well stirred waters. For example, the Middle Jurassic Great Oolite Series of southern England is characterised by cross-stratified, rippled and shelly oolitic limestones deposited in a marine intertidal flat and tidal channel situation. The bulk of the beds are oosparites or oolitic grain-stones. Oosparites are also well developed within Lower Carboniferous marine shelf successions of South Wales and England.

Pisoliths are rounded, sometimes irregular shaped bodies, greater than two millimetres in maximum diameter, which superficially resemble ooliths, yet have more complex origins. Modern pisoliths usually consist internally of alternate layers of grayish, organic-rich micrite and clear, microcrystalline aragonite, the latter showing radial orientation. These layers are wrapped around nuclei of molluscs, algae and lithoclasts. Certain varieties, recorded from the sabkha zones around the Persian Gulf, appear to be concretionary

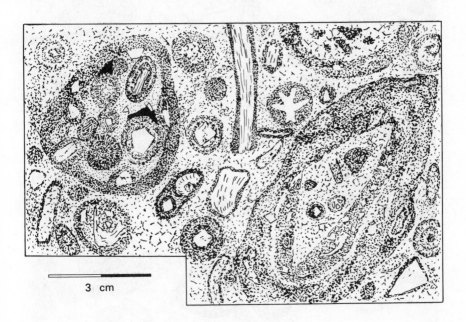

3 cm

Figure 8.15 Pisolitic limestone, Middle Jurassic, Gloucester, England. Large oolitoid bodies but much more variable in shape than normal ooliths, which are sprinkled through the sparite-micrite matrix. The pisoliths are commonly micritised around their periphery. Their internal structure is sometimes complicated and resembles that of modern grapestones. Polarised light.

in origin, the carbonate being precipitated from evaporating pore water. In the same region aragonitic, crusty **beach-rock**, on the higher levels of the tidal zone, is partly aggregated of pisolitic grains, apparently formed by direct chemical precipitation from sea-water during very high tides. Yet other types of pisoliths are forming by direct chemical precipitation in the adjacent lagoons.

Good examples of calcite pisoliths are found in the Middle Jurassic oolitic beds of England (Fig. 8.15).

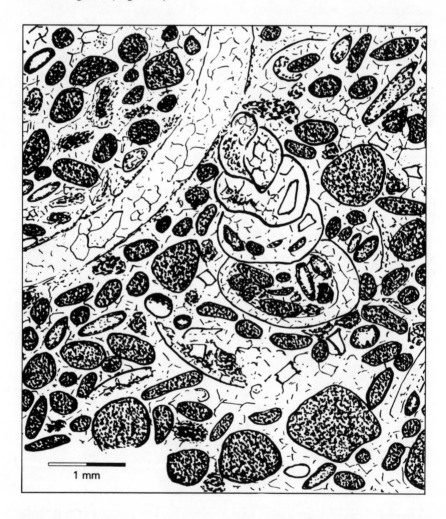

1 mm

Figure 8.16 Pelsparite, Middle Jurassic, Oxfordshire, England. The pellets are devoid of organised structure internally and are composed of micrite. They partly occupy the chambers of a gastropod shell (centre). The cement is sparite. Polarised light.

Pellet-bearing deposits

The pellet-bearing carbonate sediments (e.g. pelsparites, pelmicrites, peloid packstones) generally owe their peculiar characteristics to organic agencies. When the pellets have no detectable organic affinities the term peloid is preferable. Pellets or peloids can vary considerably in size, but usually consist of ovoid or elongate bodies less than a millimetre in size, composed of micrite and aggregated to form a rock superficially resembling an oolitic limestone (Fig. 8.16). They accumulate in quantity in low to moderate energy situations, such as on wide tidal flats and in adjacent lagoons. Foraminiferal and molluscan débris are frequent associates. The majority of the pellets are faecal in origin and formed by the activities of gastropods, worms, molluscs and crabs. The various mud-feeding invertebrates pass the sediment through their bodies and leave the rejected material in the form of subspherical, ellipsoidal and rod-shaped grains. Within any given layer faecal pellets usually show a marked uniformity in size, ranging in different rocks from 0·03 to 0·16 mm. In some lagoons of the West Indies the rod-shaped pellets of the marine gastropod *Batillaria* locally form more than 90 per cent of the bottom sediment. They are composed of silt-sized particles embedded in organic mucus with traces of algae. Both modern and fossil pellets are generally rich in organic matter which gives them a brownish tinge.

Some shallow water muds on the Florida coast contain 50 per cent faecal pellets but a metre or so below the surface it has been noted that compaction has destroyed their individuality. The total destruction of pellets by compaction is normally prevented by interstitial aragonite tending to harden the muds relatively quickly. For instance, around Andros Island the faecal pellets, which form up to 30 per cent of some sediments, are rapidly bound together quite firmly by aragonite cement.

Many fossil carbonate rocks show a pellet structure when examined in thin-section and in most cases a faecal origin seems likely. In Britain this type of lithology is seen in the Lower Carboniferous beds of the classic Avon Gorge section at Bristol, also in Wales, the northern Pennines of England and in Ireland.

ORTHOCHEMICAL LIMESTONES

These are microcrystalline sediments characterised by a marked dominance of micrite or lime mud. Lithified equivalents are referred to as calcilutites, micrites, mudstones and, occasionally, wackestones (the last two in the strict sense of Dunham only). In Recent deposits the individual aragonite particles can only be detected with the electron microscope whereas, in ancient rocks, the particles are often diagenetically enlarged, such that the granularity can be more readily observed with an ordinary microscope (Fig. 8.17).

Figure 8.17 Mudstone, Lower Carboniferous, south-west England. An even-grained rock originally formed of micrite, but subsequently modified by neomorphic enlargement into something approaching a calcisiltite. Polarised light, × 100.

Allochems of all types form a low volumetric percentage or may be completely absent. Sparry calcite tends to occur only if the original ooze has been disturbed by bottom currents, escaping gas bubbles or by the activities of organisms, especially algae. The structure created by any of these disturbances is known as 'birds-eye' and it is typical of intertidal and supratidal deposits. Folk describes these disturbed muds as **dismicrites**.

Fine-grained carbonate mud accumulation occurs in low energy environments, where there is a general lack of strong winnowing currents. Protected shallow water lagoons and shelves, and deeper off-shore basins are common situations. Deeper water varieties of mud occupy depressions at water depths greater than thirty metres in the Persian Gulf. In the Tongue of the Ocean trough in the Bahamas (Fig. 8.3) similar muds are found, some of which are laminated and show small-scale graded bedding. Turbidity current emplacement is implied, and the resultant turbidite deposits are sometimes called **allodapic**.

The best described, though not best understood, lime mud deposits are those originating in shallow waters. Major problems arise over the source of this mud. Many petrologists suggest a physicochemical origin whereas others believe a biochemical or even biogenic origin is more likely. The problem is complicated by the fact that conditions which favour one origin also favour the others. These conditions are warm surface waters saturated with $CaCO_3$, restricted circulation of the waters and a tropical or subtropical climate.

Conditions which favour all the processes of accumulation (physico-chemical, biochemical, biogenic) are well illustrated on the Great Bahama Bank, where a considerable area of the sea floor west of Andros Island is covered with a white deposit of aragonite mud. The sea floor here stands only about six metres below water level, and maintains this elevation more or less uniformly right up to the edge of the shoals, some 80 km off-shore. There is then a sudden drop, usually at a gradient of one in three, to depths varying between 600 and 1800 m. The extreme shallowness of the sea over these shoals effectively reduces circulation, and the water in the centre of the bank is virtually isolated from that of the open ocean. In the strong sub-tropical sun the water rises in temperature to as much as 31 °C in summer, and increased evaporation takes place; since the rainfall is insufficient to make up for the loss by evaporation, water is drawn in from the ocean, and there is a concentration of salts ($>41\%_0$ salinity) in the centre of the bank. The surface water of the open ocean in these latitudes is already saturated with respect to calcium carbonate, and any rise in temperature reduces the solubility of the salt; consequently, as the solution becomes warmed and concentrated in its passage across the shoals, the excess calcium carbonate crystallises out. In quiet and sheltered parts of the bank, the precipitate takes the form of minute needles of aragonite, usually 1–5 microns in length, forming a white, muddy-textured sediment of almost pure calcium carbonate. Such a sediment is known as **aragonite mud**. There is a variable percentage of skeletal fragments and pellets distributed through the mud, so the names biomicrite and pelmicrite are also appropriate (Fig. 8.3).

It is difficult on the bank to quantify the proportion of aragonite directly precipitated from sea-water and estimates vary from virtually zero to substantial amounts, greater than 50 per cent. The contribution of identifiable skeletal débris to the Bahamas muds is about 10 per cent, and the rest of the aragonite is either a chemical precipitate or unidentifiable organic débris, or a mixture of both. The proponents of biogenic sources for the mud favour calcareous algae, the argument used being the ease with which certain varieties (*Penicillus, Udotea, Rhipocephalus, Acetabularia*) disintegrate on death into single aragonite needles. Theoretical calculations show that if the average density of such algal growth reaches two to three plants per square metre, over a sufficiently large area of sea floor, then the rate of needle accumulation can approach the known rates of sediment deposition. However, until some uncontroversial method of distinguishing between organic and inorganic needles becomes available, the arguments are likely to continue.

Aragonite needles are quickly resuspended by bottom disturbance to form turbid clouds or patches, known as **whitings**. Whitings can be several hundreds of metres long and are carried along current paths until the velocity and turbulence diminish, at which point resedimentation occurs.

The drifting of needles inshore leads to a progressive accretion of aragonite mud in the intertidal and supratidal marsh zones. Whether further accessions of chemically precipitated aragonite occur in the upper tidal zones as a consequence of in-place bacterial reactions is still conjectural.

Fossilised carbonate muds are most often found in shelf successions. In Britain, some of the best known examples are again of Carboniferous age. The Cementstone Group limestones of Scotland and northern England are predominantly calcite mudstones, deposited under sabkha-lagoonal conditions, and later partly dolomitised.

The purer calcite mudstones or calcilutites are pale gray in colour, and have a smooth, almost conchoidal fracture, but small quantities of impurities may impart darker or more brilliant colours. Thin-sections show that most of these rocks consist of grains ranging between 0·5 and 4 microns in size. However, silt-size grains (4–60 microns) in patches and lenses may be present and occasionally dominant. Running through the rocks are irregularly branching little veins of sparry calcite; these are probably formed at the time of conversion from aragonite mud to a solid calcite rock. In the conversion of the aragonite all traces of needle-shaped grains are lost and the new calcite grains rapidly become cemented into a rigid non-porous fabric by marginal syntaxial outgrowths which progressively infill the intergranular pores. Pressure solution at the contacts of the calcite grains probably plays an important role in the deposition of the syntaxial outgrowths.

TERRESTRIAL DEPOSITS

Almost all natural waters contain calcium carbonate in solution. The solubility of carbon dioxide in water is increased by pressure, so that springs issuing from a deep-seated source, often at a temperature above that of the air, immediately become saturated and deposit their excess of calcium carbonate in the solid form; thus many hot springs, especially in volcanic regions, give rise to great masses of calc-sinter, **travertine** or calc-tufa, as in Auvergne, and on a much larger scale in Italy, especially in Tuscany and near Rome. The deposits shows a remarkable development of concentric and botryoidal structures, resembling those of oolitic and pisolitic limestones; while in some cases nearly spherical concretionary masses have been formed with a diameter of two or three metres, and consist of innumerable, thin concentric layers of calcium carbonate.

Perhaps the best known of all deposits are those so extensively represented in limestone caves. The water percolating through the overlying limestone becomes charged with calcium carbonate, which is deposited when the carbon dioxide is given off in contact with the air. The process is assisted by the evaporation of the water. A film of calcium carbonate forms over drops hanging from the roof, and, by gradual accretion long, pendent icicle-like bodies with radial or concentric structure are produced. Similarly, the drops falling on the floor build up masses which often take the form of pillars. The pendent masses are called **stalactites**, the floor material, **stalagmites**. Deposits of this kind occur in nearly all limestone caves.

AUTOCHTHONOUS REEF ROCKS

A well known type of lime deposition is that associated with reefs in shallow and warm waters. Reefs are the product of actively building and sediment-binding organic constituents. These constituents, mainly calcareous algae, corals, sponges and bryozoa, erect a rigid, wave-resistant structure which stands up as a definite topographic feature above the immediate areas of deposition. They have to be distinguished from banks created by the accumulation of organisms which have no sediment-binding potential. Petrographically, reef limestones can be referred to as **biolithites** or **boundstones** with suitable qualifying adjectival prefixes such as algal or coral, depending on the preponderance of any particular component.

The evidence favouring wave resistance is found around the reef proper and consists of allochemical limestones formed from abraded reef material dipping away from the reef core. On the seaward or fore-reef side talus deposits are usually thick, steeply dipping and grade into deeper water sediments. On the landward or back-reef side the reef detritus usually forms a thinner veneer and is closely intermingled with shallow water lagoonal deposits. The association of the reef proper and the sediments derived from its abrasion is commonly referred to as the **reef complex**.

If growth of reefs is extensive then their influence on sedimentation may be quite considerable. Circulation may be so restricted in the back-reef lagoonal areas that evaporites are precipitated. Terrigenous silts and muds may be prevented from migrating towards the deeper basins or may be channelled through gaps in the reef complex. In the latter case the materials may be transported via the mechanism of turbidity currents.

Modern reefs are generally confined to shallow water, but the majority of them stand at or near the edges of steep submarine slopes, so that their débris is often deposited in deep water. Reefs situated entirely in shallow water areas, such as the Alacran Reef, are of less importance at the present day than those facing deep water, but the fossil reefs of the Mesozoic and Palaeozoic limestones appear to have grown principally in areas of shallow epicontinental seas.

Modern reef rock does not consist by any means exclusively of frame-building coral – there are always present numerous shells and skeletons of marine invertebrates which live in the shelter of the reef. Encrusting algae are also important rock builders in such an environment. Nullipores (Corallinaceae, such as *Lithophyllum*) live in great abundance on the outermost part of the reef, where it is most fully exposed to wave action. Robust corals and molluscs also inhabit the more rigorous environments. *Halimeda, Penicillus,* and other green algae colonise the more sheltered parts, such as the lagoon side.

The coarse fragmental deposits associated with reefs consist of large pieces of detached coral, sometimes encrusted with *Lithophyllum*. The finer fragmental deposits are shell, coral, and foraminiferal calcarenites, and calcilutites. Not uncommonly the lagoon deposits of atolls contain large quantities of *Halimeda* segments, foraminiferal shells, and micrite with coral fragments.

The limestone of oceanic reefs is often remarkably pure, and consists almost entirely of calcite and aragonite, with practically no terrigenous minerals.

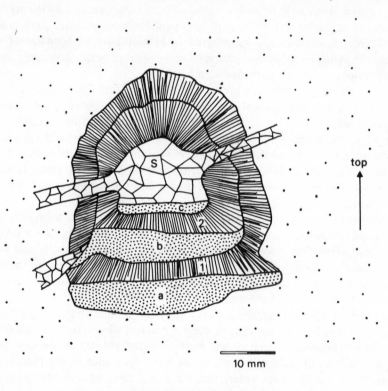

top

10 mm

a, b, c   3 generations of internal sediment
1, 2   2 generations of fibrous sparite
S   cavity fill of granular sparite

Figure 8.18 Geopetal cavity infilling in limestones. Complex chronology of micritic internal sedimentation by infiltration, followed by sparite cementation, in an algal reef of Devonian age in New South Wales. As the micritic sediment (a, b and c) usually occurs at the bottom of the cavities it is possible to deduce the original orientation of ancient folded beds.

The magnesium content is normally distinctly higher than in pure coral skeletons, which are almost non-magnesian. It is certain that much of this magnesium may be accounted for as an original constituent of the sediment, introduced in the skeletons of such organisms as foraminifera, alcyonarian corals, echinoderms and coralline algae.

A particularly interesting feature of true reef structures is a **geopetal** fabric, caused by the partial infilling of cavities by carbonate detritus (Fig. 8.18). The detritus is commonly of silt or mud grade and partially or completely infills the cavity. When incompletely filled the residual space left above the horizontally deposited silt or mud may be subsequently infilled with sparry calcite. Both the detrital and sparry infill are believed to be deposited early in the history of a reef and, clearly, they help to reduce the initial porosity quite considerably.

DIAGENESIS

Compaction and cementation processes and effects are so closely interwoven in carbonate deposits, that it is difficult to separate one from the other. The problems originate from the ease with which soluble carbonates are sedimented, dissolved and reprecipitated. Aragonite muds become slightly lithified by subaerial desiccation, and by the progressive conversion of aragonite into calcite, soon after burial. Cement is commonly introduced early into deposits, so inducing further rigidification, as in hardgrounds and beach-rocks. Under these circumstances, when original porosities of 70 per cent may be reduced quickly to less than 5 per cent, the evidence for compaction is negligible. On the other hand, if it happens that recrystallisation and cementation occur at relatively slow rates, then a certain amount of grain repacking under load is feasible, with volume reduction. Shells may show breakage and faecal pellets, ooliths and pisoliths may develop a spastolithic structure.

The development of stylolitic and microstylolitic junctions within limestones is indicative of compaction, though the depth at which the effect is generated or the reasons why some limestones are more susceptible to their formation than others are unknown. The sutured junctions, sometimes several centimetres in amplitude and tens of metres long, are caused by pressure solution mechanisms. Loading gives stress at grain contacts, solution ensues at the contacts and the carbonate taken into solution becomes available for reprecipitation as cement in adjacent pores. Suggestions have been made that pressure solution processes can reduce the thickness of partly cemented limestones by 30–40 per cent, but these figures are probably excessive.

Various aspects of the complicated subject of carbonate cementation are mentioned earlier in this chapter. As the subject is still in a very speculative state detailed elaboration on current views is best left to more advanced texts. Only a few generalities are discussed here.

At present, cementation processes are active in a wide range of environments, extending from subaerial situations of dripstone and caliche accumulation, via intertidal zones where cementation is rife, to deep sea situations of calcilutite ooze deposition. In all these environments the bulk of the cementation is concurrent with deposition and, if not, invariably early diagenetic. It is unlikely, however, that cementation ceases once the sediment is removed from contact with the depositional waters, and further accessions of cement would be expected under high loading, thick overburden pressures.

Cementation is not always a continuous, steady effect. Discontinuities are recognisable by changes in the chemical composition and morphology of the carbonate grains, or by cements clearly pre-dating and post-dating certain events, such as a phase of fracturing of shell fragments by compaction. The time intervals between cementation episodes are very variable extending from a few days to several tens of millions of years. When prolonged intervals of time are involved, sea level changes may be an important factor. For example, a sediment partly cemented in an intertidal environment may become isolated from the sea by a fall in level and quickly become susceptible to a different style of fresh and brackish water cementation.

The sources of primary cements are always a bone of contention, some authorities laying stress on direct precipitation from sea water, others on the solution of aragonitic organisms at higher levels within a carbonate layer and reprecipitation as calcite at lower levels. Pressure–solution mechanisms are considered important during the later stages of lithification. In some caliche and calcrete deposits the source is totally uncertain.

It has been claimed that the type of primary cementation can be broadly indicative of environment. In the meteoric water zone, micritic and microrhombic calcite are commonly precipitated near the ground surface and sparry calcite at greater depths. In intertidal and subtidal zones, sometimes marked by great fluctuations in salinity, cementation is generally accomplished by acicular and radial fibrous micritic aragonite and magnesian calcite.

Whether the distinctions between the cements are retained as the rock progressively lithifies is another debatable issue. Primary cements are very susceptible to change, either by removal or by in-place alteration. One common type of in-place alteration is **neomorphic aggradation**, which produces grains larger than the original, and another type is **neomorphic degradation**. which is a grain diminution process. Whenever these processes become operative it means that the fabric of the sediment, and often its mineralogy, alters; occasionally the alteration is drastic. Although there are some criteria which allow neomorphic changes to be recognised in thin-section, such as transecting boundaries, it is an inescapable fact that the post-depositional changes introduce great subjectivity into the interpretation of the origin and

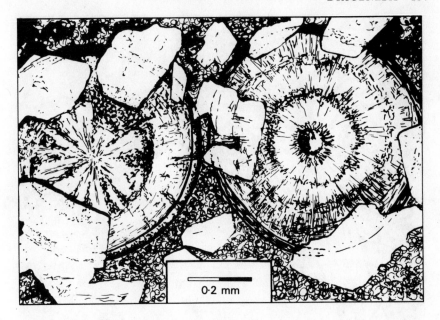

Figure 8.19   Feldspathised oolitic limestone, Precambrian, Montana. The ooliths and fine-grained matrix are partly replaced by authigenic crystals of clear orthoclase. The original concentric oolith structure can be traced through the feldspar. Polarised light.

subsequent evolution of a limestone. Perhaps it is advisable when examining limestones of all ages to assume that, unless there is strongly contradictory evidence, the rock fabric differs to some degree from that of the sediment originally laid down.

In very pure calcareous deposits such as aragonite muds, very fine particles of volcanic glass are sometimes the principal non-calcareous impurities. This observation is of considerable significance in connexion with the frequent occurrence of authigenic quartz and feldspar in some of the purer limestones, well formed crystals constituting the bulk of the insoluble residue from some of these rocks. Orthoclase is abundant in certain of the Precambrian dolomites and limestones of the Rocky Mountains in Montana, in extreme cases accounting for about 40 per cent of the rock (Fig. 8.19). The crystals are anhedral when crowded, but in many examples are well formed, and have clearly grown within the limestone, since they are found replacing original structures such as ooliths. More usually the authigenic feldspar is found to be albite, as in the Precambrian Cuddapah Limestone of Bengal and Lower Cretaceous limestones of Bavaria (Fig. 8.20).

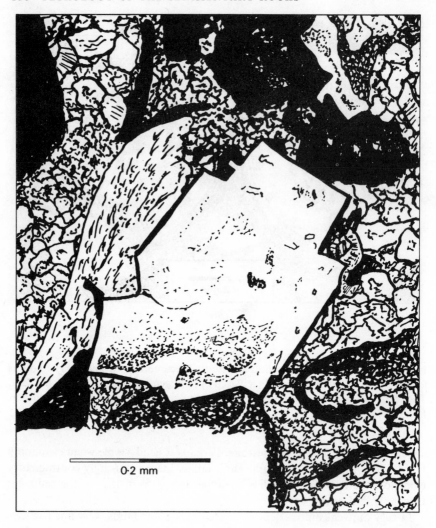

0·2 mm

Figure 8.20 Authigenic albite, Lower Cretaceous, Bavaria. The albite is growing at the expense of the shell fragments and sparry calcite matrix in the limestone. Polarised light.

The silicification of limestones, in which the carbonate of calcium is replaced by silica, generally in the form of chert, is very common. The silica is commonly derived from organic remains (sponge spicules, the frustra of diatoms, radiolaria, etc.) in the limestones, and the process consists in its solution, segregation, and redeposition. Silicified oolites are familiar examples (Fig. 8.21).

In many calcareous rocks, which have been partially silicified, the fossils only are found to be replaced, and the bulk of the rock has not been attacked. Where silicification has been more intense, the matrix of the rock is first attacked, giving rise to chert or flint. In this case the fossils, especially those with stout calcitic shells, are much more resistant to alteration than the matrix. In many of the crinoidal cherts of the Carboniferous Yoredale beds of northern England, for example, unaltered crinoid fragments are distributed throughout the chert bands or nodules with exactly the same arrangement as that to be seen in the unsilicified limestone. The canals in the crinoid ossicles within the Yoredale cherts are completely occupied by calcite, and not silica; this point is significant, for it shows that the cementation of the sediment had already started before silicification set in.

A similar conclusion is to be reached from a study of dolomitic cherts. In the Cambrian Durness Limestone of north-west Scotland for example,

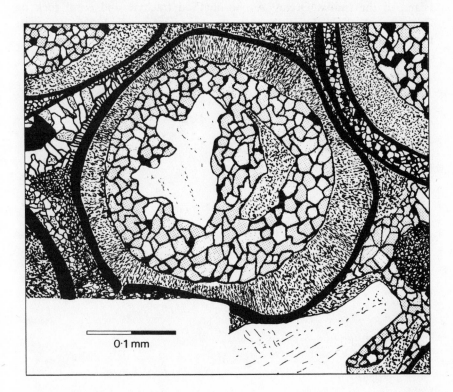

0·1 mm

Figure 8.21 Silicified oolitic limestone, Cambrian, Pennsylvania. The original limestone has been completely replaced by microcrystalline quartz and fibrous chalcedonic silica. The large quartz grain in the centre, though modified during recrystallisation, may be an original clastic grain. Crossed polars.

silicification has selectively avoided the dolomite crystals in slightly dolo-mitised limestones. The minute dolomite rhombs are very conspicuous in the chert beds, and their distribution in the unsilicified limestone may be rendered apparent by dissolving away their calcite matrix. Silicification took place only after a certain degree of dolomitisation had already affected the rock. In other examples silicification apparently took place whilst dolomitisation was still in progress.

Limestones which have been penetrated by waters carrying phosphates leached from overlying guano beds, are often highly phosphatised by metaso-matic replacement along certain zones or in isolated patches. Where the phosphate-bearing solutions attack limestones, calcium phosphates are pro-duced; such as collophanite $(Ca_3(PO_4)_2.H_2O)$ and dahllite $(4Ca_3(PO_4)_2.2CaCO_3.H_2O)$.

Examples of limestones phosphatised by meteoric water are found on the islands of Sombrero, Mona, and Moneta in the West Indies, and Remire Island in the Indian Ocean. A phosphatised trachyte and coral rock on Clipperton Atoll, in the northern Pacific, have both received their phosphate in the form of an ammonium salt by leaching from guano.

9

Magnesian limestones and dolomites

Most carbonate rocks contain magnesium carbonate in addition to calcite and aragonite, and analyses show every gradation in composition from almost pure calcium carbonate rocks to those with more than 50 per cent of magnesium carbonate (Table 9.1). The nature of the magnesium-bearing minerals, and their distribution within the rock, varies to some extent in different types of magnesian deposits. Magnesite is rare in the sedimentary rocks as an independent mineral, and is only to be found in exceptional rocks such as magnesitic dolomites and certain chemical precipitates. In the types with approximately equal proportions of magnesium and calcium, the double carbonate, dolomite, $CaCO_3.MgCO_3$, accounts for most of the magnesium present. In the less magnesian rocks, especially those containing less than 8 per cent of magnesium carbonate, this compound may either be held entirely in solid solution in calcite crystals, or may be present as isolated crystals of dolomite. The former condition is found in many biogenic limestones, the latter in rocks which have been slight metasomatised by magnesian solutions.

Because of the complexities in mineralogy the classification of these deposits on a mineralogical or chemical basis is difficult. The simplest, and most commonly used, criterion is the relative percentage of calcite to dolomite (Fig. 9.1).

Another method of tackling the classification problem is to adopt the petrographic approach of Folk. Partially dolomitised limestones are described using his shorthand nomenclature plus suitable qualifiers such as dolomitic or dolomitised (Fig. 8.2). Under this scheme, a partly dolomitised oolitic limestone with sparry calcite cement could be called a dolomitic oosparite. More complete replacement by dolomite with 'ghost' or relict textures would lead to an oolitic dolomite, and a completely replaced limestone with no 'ghost' texture would be called a dolomite, with suitable grain size designation, e.g. coarse-grained dolomite.

Table 9.1 Chemical analyses of dolomitised limestones.

	1	2	3	4	5	6
SiO_2	3·59	9·44	3·27	0·14	2·69	16·63
Al_2O_3	0·78	2·90	0·68	1·90	0·59	7·43
TiO_2	0·04	0·17	0·10	—	0·02	0·40
Fe_2O_3	0·66	1·43	6·31	0·49	0·08	0·28
FeO	—	—	—	0·17	1·30	6·03
MgO	20·53	17·76	15·38	20·75	19·45	10·20
CaO	29·25	26·72	28·85	30·07	29·16	21·62
MnO	0·01	0·15	0·31	0·02	0·15	0·24
Na_2O	0·06	0·13	0·15	—	0·05	0·14
K_2O	0·69	0·83	0·19	—	0·08	0·75
H_2O+	0·06	0·84	0·22	0·28	0·54	0·67
P_2O_5	0·02	0·16	0·18	—	nil	0·25
CO_2	44·49	38·23	43·31	46·02	44·52	31·10
FeS_2	0·05	0·20	0·29	nil	1·22	—
C	—	—	—	0·03	0·14	0·64

1 Cambro-Ordovician, Sutherland, Scotland.
2 Cementstone Group, Carboniferous, Scotland.
3 Carboniferous Limestone Series, Fife, Scotland.
4 Lower Magnesian Limestone, Permian, north-east England.
5 Lower Magnesian Limestone, Permian, east England.
6 Ankeritic siltstone horizon, Carboniferous, England.

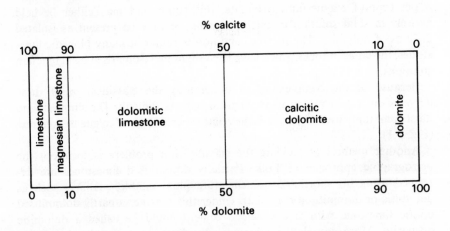

Figure 9.1 Limestone-dolomite classification.

Dololithites

These are clastic or detrital accumulations of dolomite which have been derived from pre-existing dolomitic rocks (Fig. 9.2). Subaerial weathering of mildly calcitic dolomites is capable of producing such rocks by the solution of interstitial calcite. The dolomite grains are progressively released and accumulate as sandy pockets and aprons mantling relict cores of dolomite rock.

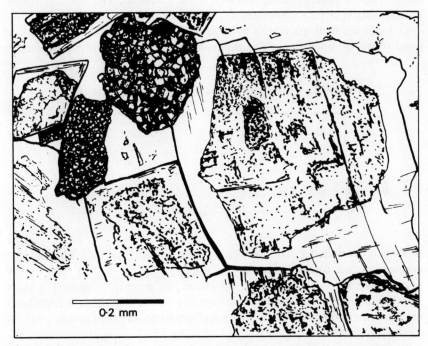

0·2 mm

Figure 9.2   Dololithite, Cretaceous, Texas. Fragments of 'dusty' detrital dolomite with syntaxial overgrowths of cleaner authigenic dolomite. Polarised light.

Certain dolomitic and calcareous sandstones consist of rounded and well sorted quartz and dolomite grains graded to approximately the same dimensions.

A very slight degree of recrystallisation is probably sufficient to destroy the clastic appearance of dololithites, and in the finer textured types the changes involved in ordinary lithification would effectively remove all microscopic evidence of their origin.

DOLOMITISATION

Most, if not all, dolomitic rocks were originally laid down as limestones, and have acquired their present composition as a result of early or late metasomatic alteration. However, among the lithified rocks a few instances have been described in which dolomite may have formed as an original precipitate. In most of these cases the deposits of dolomite are associated with rock-salt and gypsum or anhydrite, or they show other evidence of having been formed in salt lakes or enclosed seas.

Dolomite rocks preserve the structures and textures of their parents to varying degrees. Dolomite has a very strong tendency to form idiomorphic crystals, even when growing within a solid limestone, and for this reason dolomitisation largely obliterates any primary structures.

Dolomitisation is often seen to be selective in its attack upon the various constituents of limestones. The dolomitisation of ooliths often appears to

Table 9.2   Carbonate reservoir rocks, south-west Iran.

Dolomitisation (%)	Porosity (%)
0	0–4
20	4–8
32	8–12
Over 58	Over 12

proceed independently and in advance of the cement (Fig. 9.3). The fact that magnesium-rich calcareous algae are present in some ooliths may be important in their speedier rate of alteration. In general, aragonite is more readily altered than calcite (even though the calcite may have a high magnesium content) and fine-grained compact material more readily than coarsely crystalline structures.

Aragonitic shells such as those of gastropods and cephalopods are the first to suffer dolomitisation, and are often found to have been altered before the matrix. In contrast, stout calcitic fossils in calcarenites will often resist dolomitisation even though the enclosing sparite and micrite matrix may be completely altered. In this way are formed dolomitic rocks enclosing more or less unaltered fossils, such as crinoids, stout brachiopod shells, or rugose corals.

The process of dolomitisation itself may have more far reaching effects on porosity. The replacement of calcite by dolomite on a molecule-by-molecule basis involves a contraction in size of 12–13 per cent. Under ideal circumstances this can mean that a limestone with a rigid fabric can have its porosity increased by 10 per cent or more (Table 9.2). In some Arabian Jurassic oil reservoir limestones porosities of 19 per cent are recorded in the more highly dolomitised layers; permeability is also comparatively high.

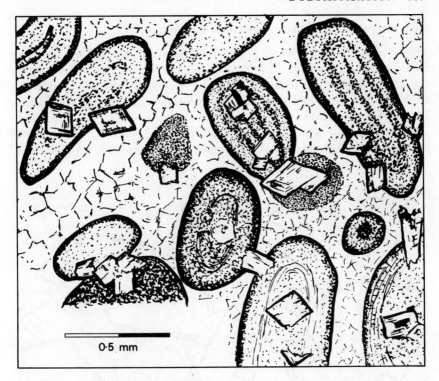

Figure 9.3 Partially dolomitised oolitic limestone, Middle Devonian, South Morocco. A dolomitised oosparite. The rhombs of dolomite show preferential generation within the ooliths, though transecting external boundaries of the allochems. Polarised light.

The actual time which elapses between the deposition of the limestone and its subsequent dolomitisation varies considerably in different cases. Many fine-grained carbonate deposits are altered very soon after deposition, whilst they are still lying in an unconsolidated condition. In other cases, there is clear evidence that dolomitisation took place long after the original rock had been lithified, the magnesian solutions finding an entry through open joints and minute pores. It is thus important to distinguish between rocks affected by early diagenetic or pene-contemporaneous dolomitisation, and those which consolidated as limestones, and later were subjected to late diagenetic and subsequent dolomitisation.

The final general point about dolomite is that weathering in the presence of pyrite or sulphate can bring about reconversion to calcite. This process of **dedolomitisation** has been noted in Recent sediments of the Middle East and the Carboniferous Limestone of the Isle of Man and the Upper Jurassic Gigas Beds of Germany.

Early diagenetic (pene-contemporaneous) dolomitisation

Most bedded dolomites which preserve an approximately uniform character over a wide area are believed to have undergone metasomatism shortly after deposition. Dolomitic rocks of this kind are sometimes of considerable thickness and form important stratigraphical units, such as the Sailmhor Dolomite (Cambrian) of north-west Scotland, the laminosa-Dolomite (Lower Carboniferous) of South Wales, the Magnesian Limestones (Permian) of north-east England and the Knox Dolomite (Ordovician) of the Appalachians of eastern North America (Fig. 9.4).

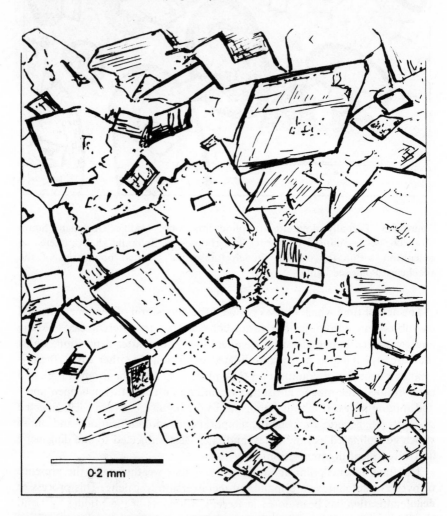

0·2 mm

Figure 9.4 Knox Dolomite, Lower Palaeozoic, Tennessee. A dense intergrowth of dolomite grains showing good rhombic shapes. Polarised light.

The evidence for the pene-contemporaneous date of alteration varies with the individual character and circumstances of each bed in question. In many cases, stratigraphical evidence is sufficient, as in those examples which lie between unaltered limestones, showing that magnesian solutions cannot have been introduced subsequently. The presence of dolomitised fossils within the formation is sufficient proof that the dolomite is not a primary deposit. In South Wales, pebbles of dolomite may be found enclosed within the overlying conformable limestone formation.

Favourable conditions for pene-contemporaneous dolomitisation are:

(a) Shallow warm water between 0 and 45 metres in depth.
(b) The presence of carbon dioxide in comparative abundance, causing partial solution of the limestones, and the possibility of chemical interchange with the magnesium salts in sea-water.
(c) Porosity of the limestones, allowing the percolation of sea-water through the mass of the rocks.
(d) Sufficiently slow subsidence or elevation to render complete the change from calcium carbonate to the double carbonate of calcium and magnesium.

To these must be added: (1) water having a relatively high Mg/Ca ratio preferably in excess of 20 to 1 (normal sea-water is 3–5 to 1); and (2) an adequate rate of production and flow of this water or brine so as to convert the sediment in the time available.

These conditions are known to be attained to varying extents on modern supratidal flats in many humid parts of the world, including the Bahamas, West Indies and Persian Gulf. Supratidal flats, known as coastal sabkhas in the Middle East, lie above normal high tide level and are only flooded by highest spring and storm tides. They are subaerially exposed for relatively long periods of time between these highest tides. As a consequence, the laminated, sun-cracked lime muds are only subject to intermittent soaking by normal sea-water. Some of the pore water almost certainly emanates by groundwater percolation from the adjacent sea. Evaporation then increases the concentration of salts in the pore waters near to the surface until the point is reached where calcium sulphate is precipitated in quantity. This fixation of calcium simultaneously removes the chemically inhibiting influence of calcium sulphate in solution on dolomite precipitation and increases the Mg/Ca ratio of the residual brines. The conditions are then appropriate for the replacement of carbonate in the original muds by dolomite. Aragonite is replaced in preference to calcite at this stage. Selective dolomitisation of aragonite pellets and aragonite-filled algal borings in skeletal calcite fragments are cited as evidence.

The proportion of original calcite in the upper layers of the sediment is usually diminished during dolomitisation and is ascribed to contemporary leaching processes. The carbonate ions taken into solution are probably immediately incorporated into the new dolomite grains.

The net result of all these complex processes is the production of very porous, light-brown to gray crusts, patches and layers of dolomite in the top part of the lime muds. Dolomite rhombs, 1–5 microns in size, form 20–95 per cent of these. On the west side of Andros Island in the Bahamas dolomite up to 150 centimetres thick has formed in the last 5000 years.

Late diagenetic and subsequent dolomitisation

Late and subsequent dolomitisation effects are often difficult to diagnose successfully because they are caused by a progressive series of events spread over what may be a considerable length of time. Late diagenetic dolomitisation by connate and groundwaters occurs under more extreme loading and thermodynamic conditions than early diagenetic changes. However, some of the deeper-seated effects are probably contemporary with pene-contemporaneous dolomitisation effects at the surface (Fig. 9.5). Some contemporary brines may seep downwards into underlying beds and create

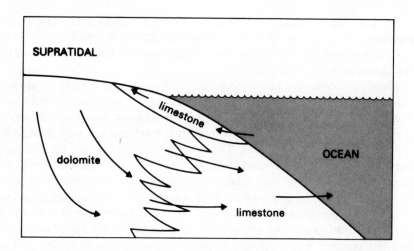

Figure 9.5 Dolomitisation and refluxion. A hypothetical diagram illustrating the possible relationships between environments, surface and near-surface influxing and deeper refluxing waters, and styles of dolomitisation. The sea-water movement into the supratidal zone is triggered by evaporation on the sabkha. The brines convert the surface of the lime deposits into early diagenetic, pene-contemporaneous dolomite. Any dense brines sinking to lower levels and seeping back towards the sea might produce very patchy, late diagenetic discordant dolomitised bodies at depth.

a whole host of cross-cutting dolomitic textures and structures. Primary bedding features may begin to be transgressed by patches of dolomite, and structures, such as ooliths, fossils and sun-cracks, may become partially obliterated. Porosity also changes by contemporary solution of calcium carbonate, though this may be modified by later precipitation of new dolomite or calcite.

Subsequent changes by circulating Mg-rich groundwaters are considered to be associated with periods of mineralisation, uplift and emergence of the rocks and they often have a regional character being found in the more highly folded and faulted zones. Shearing pressures seem to favour this late style of dolomitisation (Fig. 9.6).

Dolomites of this type are most easily recognised in formations which are not completely metasomatised, but still contain relics of unaltered limestone. In the most characteristic occurrences, the secondary dolomite is clearly related to the planes of weakness normally found in solid rocks, the commonest channels of dolomitisation being fault-planes, joints, and minor fractures.

The Carboniferous Limestone of Britain is affected by both pene-

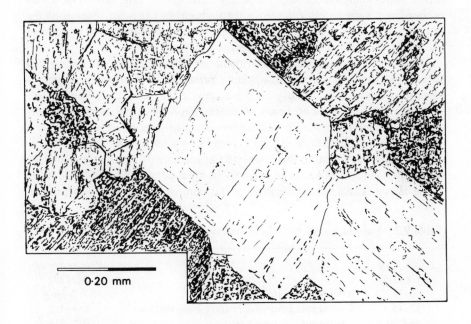

0·20 mm

Figure 9.6 Late diagenetic dolomite, Derbyshire, England. Two phases of dolomitisation of an original Lower Carboniferous limestone. The first phase is represented by 'dusty' subhedral dolomite grains and the subsequent phase by clean rhombic euhedra of dolomite. Polarised light.

contemporaneous and subsequent dolomitisation. In some districts, for example in Leicestershire, the dolomite grains are frequently well formed rhombohedra zoned with haematite, and their presence in bedded formations underlying the New Red Sandstone is taken as conclusive evidence that the rock was dolomitised in Triassic times. The gray and yellow dolomites of neighbouring areas, on the other hand, contain no zonal inclusions of haematite and the alteration of these rocks is regarded as pene-contemporaneous.

Dolomitisation of Recent and Tertiary coral reefs

In many coral islands in the Pacific and Indian oceans where the climate promotes increased salinity by evaporation, the reefs and associated organic limestones are found to have been converted partially or wholly to dolomite. Examination of raised reefs shows that many of these are also dolomitised, and certain bore holes which have been sunk into one or two islands have also encountered dolomitic rock. The foundations of the atoll of Funafuti, in the south-western Pacific, have been investigated in great detail from this point of view. A bore hole was sunk to a depth of 330 metres, and the cores raised were analysed at approximately three-metre intervals. The rocks encountered were purely calcareous throughout, but the distribution of the minerals aragonite, calcite, and dolomite showed remarkable and suggestive variations. This distribution is summarised in Table 9.3.

Aragonite is an important constituent in the uppermost 30 metres but below that depth it rapidly disappears, principally by solution, leaving a cavernous calcite rock. Nearly all the aragonite has gone at 45 metres, and no traces at all were found below 66 metres. From 66 to 190 metres the limestone is extremely incoherent and cavernous, consisting entirely of calcite in the form of organic fragments, and resembling a loose sand of reef débris. No dolomite was found in the upper part of the core (0–190 metres), and the magnesium carbonate found in these rocks must be present in solid solution in the calcite. The actual percentage of this compound varies considerably at

Table 9.3 Funafuti atoll bore hole succession.

Depth from surface (m)	Nature of Rock	Percentage of $MgCO_3$
0–30	Aragonite-calcite-limestone	3·2–16·4
30–45	Transition: aragonite progressively dissolved	2·1–4·6
45–190	Cavernous calcite-limestone	0·8–5·4
190–195	Transition: dolomite appears progressively	20·4–38·4
195–330	Dolomite rock (calcite exceptionally present)	40 (average)

different levels and it seems probable that these fluctuations depend upon changes in the proportions and nature of the magnesian organisms which have contributed to the formation of the limestone. Crystals of dolomite first appear at 191 metres, and within a short distance the core acquires the character of a dolomite rock in which independent crystals of calcite cannot, as a rule, be detected. At two or three restricted levels this dolomite rock was found to be interrupted by beds of dolomitic limestone, but with these exceptions it maintains its character to the greatest depth reached at 330 metres.

Dolomitisation probably started early in the growth of the atoll prior to the removal of the aragonite and some of the aragonite may well have been directly replaced by dolomite. Residual aragonite may have been leached in the 30–190 metres levels during the late-Pleistocene emergence of the atoll after the dolomitisation. If this is so, then the upper 30–45 metres of reef growth where aragonite exists is probably post-Pleistocene in age. Where dolomite first appears in the samples examined, its crystals are added to the framework of the cavernous calcite limestone, and clearly have grown upon the surfaces of existing calcite crystals. The dolomite is possibly a direct precipitate from solution but it could also have formed by partial replacement of high-magnesian micrite muds which originally occupied the cavities. These muds, of which traces remain, are derived largely from algae and forams.

It will be evident from this account that the course of dolomitisation in an existing reef area such as Funafuti is very complex and open to much controversy. Similar complexities should be expected in more ancient reefs.

MAGNESITIC DEPOSITS

Sedimentary rocks containing free magnesite as a principal constituent are much less abundant than calcitic or dolomitic limestones. Most occurrences of magnesite either result from the weathering of olivine-bearing ultrabasic rocks, or are formed by the action of magnesium-bearing solutions on other carbonate rocks, particularly limestones. At present magnesite is forming in supratidal muds of the Persian Gulf.

Primary precipitates of magnesite are distinctly rare, and the few known examples have been formed, not in a marine environment, but in continental salt lakes. The deposits near Bissell, in California, appear to have accumulated in such a lake. The magnesite forms a thin-bedded, compact white rock, interstratified with gray or green clays. Analyses indicate the presence of basic carbonates such as hydromagnesite, $3MgCO_3.Mg(OH)_2.3H_2O$, in addition to normal magnesite. These deposits were evidently formed in a salt lake lying on the floor of a desert basin, and it has been suggested that the lake water contained sodium carbonate, which would react with the magnesium sulphate solutions brought in by spring water, precipitating hydromagnesite at the point of entry.

Certain coastal lakes of the arid and semi-arid parts of South Australia are characterised by the seasonal (summer) deposition of hydromagnesite and magnesite in association with dolomite. The influence of algae in producing high pH values in the water and concentrating magnesium in their skeletons may be a key factor in magnesite deposition, and it has been observed that certain major fossil deposits of magnesite are intimately associated with biohermal algal dolomites. In Pitharagurh, India, lenticular magnesite beds, extending some 130 kilometres at outcrop and having thicknesses up to 30 metres, show this association. The beds contain about 42 per cent MgO and $1 \cdot 5$ per cent CaO.

10
Siliceous deposits

These include purely organic deposits in which the organic remains are embedded in a matrix or cement of minutely crystalline silica. There is no sharp line of distinction between these deposits and some varieties of chert which bear the characteristics of metasomatic replacements of limestones. Between the two extremes of purely organic and purely metasomatic types there are innumerable intermediate varieties in which various proportions of siliceous organic remains can be recognised. Certain siliceous rocks have been claimed as chemical deposits derived from primary inorganic precipitates of colloidal silica. They include extensive bedded and nodular cherts of Pleistocene and Precambrian ages. In the Pleistocene alkaline lakes of the Magadi Basin of Kenya, silica appears to have been precipitated as $NaSi_7O_{13}(OH)_3.3H_2O$ (magadiite) initially, but this mineral was progressively leached of sodium and converted into chert. Precambrian iron formations partly consist of chert for which there is only meagre evidence of biologic influence during deposition. Yet other varieties of chert may have formed diagenetically from volcanic ash laid down in water.

The terminology of the siliceous rocks requires a few words of explanation. The soft, entirely unconsolidated organic deposits are called **oozes**. Deposits which have remained unconsolidated although no longer in process of accumulation are termed **earths**, such as **radiolarian earths** and **diatomaceous earths**. The consolidated equivalents of these purely organic accumulations are termed **radiolarites** and **diatomites**; to these may be added the **sponge-spicule rocks**, which have no extensively developed equivalent amongst the unconsolidated sediments.

The name **chert** is used for all siliceous deposits of a sedimentary nature whose main constituent is redistributed silica. Thus the cherts include rocks which are organic or inorganic in their immediate origin; and those radiolarites and diatomites which have a well developed siliceous cement or ground mass may equally well be termed radiolarian cherts and diatomaceous cherts. In general, distinctive kinds of cherts are designated in this manner by the use of some qualifying descriptive term, but a few individual names, such as jasper, novaculite, lydite (lydian stone), and flint, are in common use for special varieties of cherty rocks. **Porcellanites**, though not true cherts, are dull-looking, porous and highly siliceous sediments. They are widespread in deep sea successions of late-Cretaceous to Pliocene age and are composed of low temperature cristobalite (SiO_2), mainly of biogenic origin. Representative analyses of siliceous rocks are listed in Table 10.1.

Table 10.1 Chemical analyses of siliceous deposits.

	1	2	3	4	5	6	7	8
SiO_2	60·14	73·16	66·67	95·00	61·42	98·36	94·72	94·60
Al_2O_3	13·01	9·08	14·82	1·70	0·10	0·12	0·55	1·80
TiO_2	0·53	0·21	0·63	0·08	—	—	2·13	0·07
Fe_2O_3	5·96	1·44	6·51	0·78	30·64	—	1·74	0·36
FeO	—	—	—	0·11	3·23	0·08	—	0·28
MgO	2·34	0·23	2·58	0·23	0·20	0·01	0·23	0·64
CaO	1·87	0·50	2·26	0·41	0·48	0·16	0·63	0·49
MnO	0·74	—	0·28	0·18	—	—	—	—
Na_2O	1·97	1·20	1·66	0·09 ⎫ 0·94	0·04	—	0·15
K_2O	2·04	1·40	2·54	0·33 ⎭	0·04	—	0·37
H_2O+	8·01	—	—	0·71	0·08	0·84	—	0·77
H_2O-	—	7·20	—	0·08	1·79	0·11	—	0·09
P_2O_5	0·12	—	0·14	0·20	—	0·02	—	0·24
CO_2	—	—	0·51	0·06	0·66	0·07	—	0·60
C	—	—	1·32	—	—	0·02	—	—

1 Diatom ooze, north Pacific.
2 Diatomite (lacustrine), Pleistocene, New Zealand.
3 Radiolarian ooze, north Pacific.
4 Radiolarian chert, Jurassic, California.
5 Chert band in ironstone-formation, Precambrian, India.
6 Flint, Chalk, Dover, England.
7 Silcrete, overlying Dwyka Shale, South Africa.
8 Spicular chert, Lower Permian, Nevada.

Several forms of silica are found in the siliceous deposits under consideration.

Microquartz is a non-clastic form of anhydrous crystalline silica found in sediments, and has a grain size usually less than twenty microns.

Opal is a colloidal form of silica, containing a varying proportion of water, usually between 3 and 9 per cent. Optically it differs from quartz in being isotropic, and in having a lower refractive index. The hardness and specific gravity vary with the quantity of water present, and are always distinctly lower than the corresponding values for quartz. Opal is the form found in siliceous skeletons, and in the deposits of hot springs. It also occurs as a prominent cement in some sandstones which have been subject to weathering in semi-arid regions. Opal is normally absent in cherts older than the Mesozoic.

Chalcedonic silica is the name given to the minutely granular or fibrous material present in many forms in chert. It is more stable than opal and commonly appears to have recrystallised from it. The surface of chalcedonic silica usually has a spongy appearance due to the presence of minute, spherical, water-filled cavities. It is these cavities which give the silica its brownish colour in thin-section under natural light.

SILICA PRECIPITATION

Most silica is transported into seas in true solution, probably as monosilicic acid (H_4SiO_4), and is very difficult to precipitate directly by electrolytic action. As the silica concentrations in rivers are generally higher by a factor of at least two or three than in sea-water there must be mechanisms accounting for the silica depletion. Changes in alkalinity have no effect on monosilicic acid. Possibly the silica in solution is adsorbed or coprecipitated onto colloidal and suspended solids within the river waters. Where these waters meet the sea water, with its abundant electrolytes, almost all of the soluble silica might be removed, depositing it and the solids on the sea bottom. On the other hand, considerable doubt has been cast on primary inorganic precipitation hypotheses. There is a considerable body of opinion favouring the view that most transported silica is initially utilised by micro-organisms (diatoms, radiolaria, sponges, etc.) in building their skeletons. On death, the siliceous skeletons accumulate and, if they do so in sufficient quantity, may recrystallise near to and just below the depositional interface to form cherty rocks.

In oceanic sediments the highest concentrations of dissolved silica are found in the top two metres of the bottom deposits and are predominantly due to the partial solution of siliceous planktonic débris. Siliceous pyroclastic débris supplies very little silica. In carbonate-rich sediment these conditions slowly lead to the formation of nodular chert, whereas in clay-rich sediment, porcellanite and bedded chert are formed. Cristobalite is usually an intermediate metastable product during these diagenetic changes.

Diatomaceous deposits

The diatoms are minute unicellular algae with cell walls composed of opaline silica. They form an important part of the plankton both in fresh-water lakes and in the sea. Diatoms are most abundant in the upper layers of water, being found in the sea down to depths of 135 metres, according to the penetration of light. The siliceous remains of the more robust species may be incorporated into marine sediments at any depth; however, they are not usually conspicuous in shallow water deposits, probably because they are masked by an abundance of other material. At the present day important deposits of diatomaceous ooze occur in the deep water of the polar seas and in lower-latitude, cold-current areas bounding continental masses, for example the west coast of South America. In these areas there is considerable exchange between surface waters and nutrient-rich, upwelling deep waters to produce high fertility all year round. The concentration of dissolved silica is several times greater than in surface waters elsewhere.

In the Antarctic diatomaceous ooze forms a continuous belt encircling the zone of terrigenous deposits which fringes the edge of permanent ice. Consequently the ooze often contains a considerable admixture of terrigenous material carried by floating ice. The deposit is chiefly composed of the siliceous frustules of diatoms, together with some calcareous foraminifera; other constituents of organic origin are very rare. When dried, the ooze is a white or yellowish powdery substance, which when examined under the microscope strongly resembles the diatom earth of fresh-water origin (Tripoli powder or kieselguhr).

Shallow water diatomaceous deposits are found at the present day principally in fresh-water lakes. Diatomaceous earths of Pleistocene age are widespread in Scotland and northern England, usually in the form of lenticular beds among lacustrine deposits, especially those of late-glacial and post-glacial lakes.

The diatomaceous earth of Kentmere, in the English Lake District, occupies an old lake basin of this kind. The deposit consists of diatom frustules mixed in varying proportions with clay and peaty matter, and is dark brown when fresh, but rapidly changes colour to deep olive green on exposure to the air. In its natural condition it is extremely porous.

Diatomaceous deposits of Mesozoic and Cenozoic age are known from many parts of the world. Lenticular beds, as much as thirty metres thick and devoid of internal stratification, were deposited extensively in the Rift Valley lakes of Kenya throughout Cenozoic times. Some of the beds are very pure with silica contents exceeding 80 per cent.

Radiolarian deposits

Unlike the diatoms, the radiolaria are confined to marine environments, where they tend to be most abundant in tropical rather than in polar waters. At the present day they accumulate to form **radiolarian ooze** in which the siliceous skeletons of radiolaria are a noteworthy constituent, ranging from 30 to 80 per cent (Fig. 10.1). It is usually a reddish deposit, less plastic than the brown clay, owing to the smaller proportion of argillaceous material; apart from siliceous organisms the mineral components are much the same as in brown clay. Some specimens contain an appreciable amount of carbonate of lime, chiefly in the form of foraminifera; diatoms and the spicules of siliceous sponges are also present. Radiolarian ooze appears to be confined to great depths, and specimens are dredged up from the deepest soundings

Figure 10.1 Radiolaria in decalcified Globigerina ooze. The siliceous tests were obtained from a depth of 5071 metres in the Indian Ocean. Typical radiolarian ooze consists of similar material. Magnified 400 times.

in the western Pacific and in the Indian Ocean, where it is believed to cover a total area of about 4 500 000 square kilometres.

Even the most typical radiolarian oozes can only be regarded as varieties of brown clay specially rich in siliceous organisms: both are equally deficient in carbonate of lime. However, numerous transitional forms between siliceous radiolarian ooze and calcareous globigerina ooze are known to exist. Deposits rich in radiolaria are for the most part confined to tropical and subtropical regions.

Deposits containing abundant remains of radiolaria are known in the Ordovician, and are found in most geological systems from the Lower Palaeozoic onwards to the Pleistocene. In addition, many of the cherts of Precambrian age resemble the Palaeozoic radiolarian cherts very closely, and quite well may have had a similar origin. Many of the older radiolarian deposits are known to have accumulated in shallow water; others, such as those associated with pillow lavas, were formed on the sea floor at depths which it is not always possible to evaluate, and a few examples, such as those of Barbados and Timor, may be true abyssal deposits, comparable with the modern radiolarian oozes although they are associated with globigerina marls. In Greece, thinly bedded Mesozoic radiolarian cherts appear to have formed diagenetically within the basal radiolaria-rich division of silt and red clay graded units of presumed turbidite origin. Radiolaria are by no means confined to deposits such as these, but are widely distributed in fine-textured marine sediments. Because of their small size, they are usually inconspicuous in clastic deposits, but may be readily detected in the insoluble residues of many fine-grained limestones, such as the chalks. Radiolaria are also frequently found in phosphatic nodules, often in an excellent condition of preservation.

In the more recent radiolarian deposits, such as those in the Oceanic Series (Upper Oligocene and Lower Miocene) of Barbados, the structural details of the radiolarian tests are perfectly preserved. Even in some Carboniferous deposits the preservation is sufficiently good to suggest the original sculpture of the shell. In most of the Palaeozoic cherts, however, only the shape of the radiolarian test is preserved in crystals of perfectly clear silica, surrounded by a turbid and darker matrix of minutely crystalline chalcedony.

The unlithified radiolarian deposits, known as the **radiolarian earths**, are distinctly rare, and the great majority of radiolarian sediments are comparatively hard rocks, which may be grouped together as the **radiolarites**.

The **radiolarian cherts** in Britain are principally found in the Precambrian, Ordovician, and marine Devonian, commonly accompanying volcanic outbursts, and frequently showing an intimate association with pillow lavas or greenstones derived from spilitic rocks. The cherts of the Dalradian of Scotland and the Precambrian Mona Complex of Anglesey are bright red in colour, due to the presence of finely divided ferric oxide; the cherts of Ordovician age in the south of Scotland are variously coloured, and may be gray or black, due to traces of carbon, brown or red, due to iron oxide, or greenish-gray, probably due to traces of chlorite or epidote. They are extemely fine-grained, and are characterised by the absence or extreme rarity of any detrital constituents other than slight traces of argillaceous matter, the bulk of the rock being made up of cryptocrystalline chalcedony. These rocks are extremely compact, and in most cases show little or no trace of bedding. Fossils other than radiolaria are not normally present; siliceous spicular bodies are usually to be seen, but most of these appear to be derived from spinose radiolaria, and it is doubtful to what extent sponge remains are present. The original sediment must have been similar in many respects to a modern radiolarian ooze, which suffered intense silicification probably very soon after deposition. The depth of water under which these sediments accumulated is not easy to determine, but there is no evidence of extreme shallowness.

The radiolarian cherts of Carboniferous age are of a rather different type.

They are usually well bedded or laminated sediments. Most speciments are of neutral tint, being gray, black or white according to the amount of carbon they contain. In south-west England the cherts show pene-contemporaneous crumpling effects ascribed to movement down slopes in waters considered to have been as much as 500 metres deep.

In the Alps, cherts of Jurassic age have been claimed as abyssal deposits which accumulated in the troughs of deep geosynclinal basins. Deposits of similar age and type are associated with the Zagros geosyncline of the Middle East (Fig. 10.2).

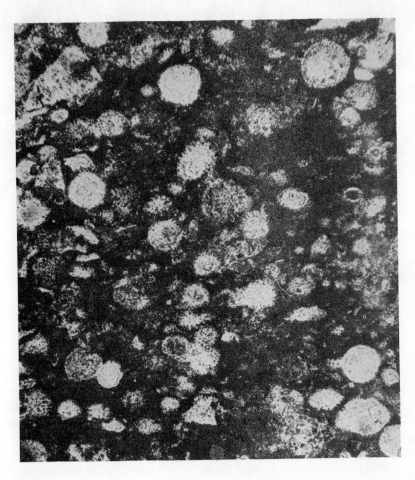

Figure 10.2 Radiolarian chert, Mesozoic, Oman. The circular objects are the remains of the radiolaria, now infilled with microcrystalline quartz. Polarised light, × 30.

Sponge spicule deposits

Although the siliceous sponges flourish principally in sea-water, they are not entirely confined to such an environment, and spicules are substantial contributors to the diatomaceous deposits of a few fresh-water lakes. Many dredgings taken from oceanic floors at present contain sponge spicules in noticeable quantities.

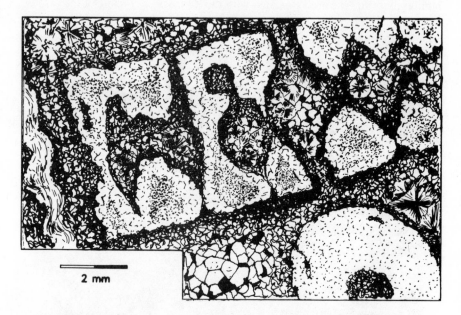

Figure 10.3 Chert, Lower Carboniferous, north England. Relict calcareous crinoid ossicles totally circumscribed and marginally replaced by fibrous chalcedonic silica and microcrystalline quartz. Crossed polars.

In Great Britain sponge remains play a variable role in the formation of cherts of Carboniferous, Jurassic and Cretaceous ages. The Carboniferous cherts form nodular bands and sheets, and vary in colour from pale gray to brown and black, depending on the composition of the enclosing limestone or shale. They are replacive in origin and it is not uncommon to find a wide range of marine organisms other than sponges showing all degrees of alteration (Fig. 10.3).

The cherts in the Purbeck Beds (Jurassic) of southern England contain spicules of the fresh water sponge *Spongilla*, and in places the chert is almost entirely made up of these spicules.

The Lower Greensand (Cretaceous) of southern England contains accumulations of spicules forming beds which vary from less than one centimetre

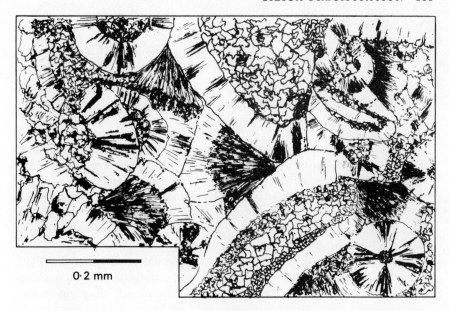

0·2 mm

Figure 10.4 Chert, Upper Cretaceous, south-west England. Two generations of fibrous chalcedonic silica, the first in the form of rounded, occasionally spherical, objects with curved outer bands, and the second mainly in the form of darker interstitial sheaves. Both types of chalcedony appear to have recrystallised early into microcrystalline quartz, which generally occupies the central parts of the rounded objects. Crossed polars.

to more than a metre in thickness. The spicules are admixed with variable amounts of detrital material including glauconite. At many localities both the Lower and Upper Greensands contain spicule-bearing cherts, though it is far from clear as to the role the spicules have played in the formation of the chert. A considerable amount of recrystallisation has occurred during the diagenesis of the cherts with, in many cases, a loss of original internal texture (Fig. 10.4).

Flint

The name flint is used principally for the siliceous nodules, nodular sheets and lenses in the White Chalk of western Europe, and in other regions where a similar facies is developed in the Upper Cretaceous. It is also used in describing the pebbly constituents of many younger detrital deposits of Tertiary and Quaternary age. The terrace gravels of the Quaternary River Thames and associated drainage systems in south-eastern England, a region of extensive Chalk outcrop, are dominated by reworked, multicyclic flints.

Flint is typically a very compact, hard substance, brittle when fresh; it breaks with a conchoidal fracture, giving clean and smooth surfaces. Thin splinters are translucent, and the colour of freshly broken nodules is usually brown, black or gray. Fresh flint has a surface patina of partly silicified chalk, usually chalky-white in appearance.

Flint consists of very fine-grained chalcedonic quartz and is generally believed to be biogenic in origin. The silica is derived from the solution of opaline silica skeletons of sponges and radiolaria, the contribution of diatoms being uncertain. Solution is initiated early in the top few metres of the chalky ooze, and the silica is redistributed and locally concentrated and precipitated as cristobalitic porcellanite, especially in bioturbated and organically-rich patches of the chalk. The concentration of decaying organic matter in these patches may well have lowered the solubility of silica, allowing pores to be filled and fossils to be replaced. Many flints are casts of burrows.

These early diagenetic processes probably account for flint bands roughly paralleling bedding, though they usually post-date pene-contemporaneous scour at the bottom of the Chalk sea. Further accessions of silica onto nuclei, and recrystallisation into chalcedonic silica, probably occurred during later diagenetic phases, with further replacement of the enveloping chalk.

11

Ferruginous deposits

The iron-bearing sediments show a considerable variety in both their mineral composition and their chemical composition (Table 11.1).

Table 11.1 Chemical analyses of ironstones.

	1	2	3	4	5	6
SiO_2	36·67	50·96	24·25	—	13·5	9·26
Al_2O_3	6·90	1·09	1·71	0·74	10·2	7·92
Fe_2O_3	—	5·01	0·71	0·53	3·0	2·80
FeO	2·35	30·37	35·22	39·87	32·5	40·54
FeS_2	38·70	—	—	—	—	—
MgO	0·65	5·26	3·16	2·64	3·5	4·43
CaO	0·13	0·04	1·78	2·12	5·0	4·75
MnO	tr	=	2·11	1·38	0·7	0·39
Na_2O	0·26	—	0·04	—	—	—
K_2O	1·81	—	0·20	—	—	—
H_2O-	0·55	0·75	0·21	0·59	8·0	—
H_2O+	1·25	6·41	—	1·21	2·9	0·30
P_2O_5	0·20	—	0·91	0·69	1·0	1·66
CO_2	—	—	27·60	28·47	19·0	27·99
Organic C	7·60	0·21	1·96	0·83	0·2	—
S	—	tr	—	tr	0·2	0·01
SO_3	2·60	—	—	—	—	—

1 Sulphide iron-formation, Huronian, Michigan.
2 Biwabik greenalite rock, Precambrian, Mesabi.
3 Banded chert-carbonate iron-formation, Huronian, Michigan.
4 Ironstone, Carboniferous, north England.
5 Main Seam, Lias, Cleveland, north-east England.
6 Sideritic chamosite oolite, Middle Jurassic, central England.

The chemical controls on iron precipitation are generally understood and reflected in a wide range of diagrams, such as Figure 11.1. From these it can

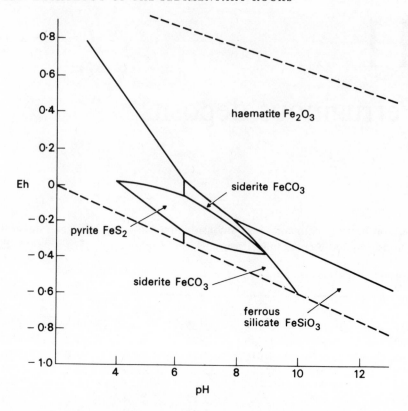

Figure 11.1 Stability fields for iron minerals. Eh–pH relationships at 25 °C and one atmosphere pressure, in the presence of water.

be deduced, excepting **magnetite** (Fe_3O_4), that Eh is much more important than pH in determining which mineral is likely to be precipitated. **Haematite** (Fe_2O_3) is precipitated and is most stable under oxidising (positive Eh) conditions, **siderite** ($FeCO_3$) will form under moderately reducing conditions, and **pyrite** (FeS_2) deposition is favoured by moderate to strong reducing conditions (negative Eh). However, environmental controls, more especially with ancient iron deposits, remain a matter for much speculation.

TRANSPORTATION AND DEPOSITION OF IRON

In underground water with a deficiency of oxygen, iron appears to be carried in the form of ferrous salts, most frequently as the carbonate, chloride and sulphate. In well aerated surface waters, however, these salts are liable to hydrolysis and oxidation, with the production of ferric hydroxide, part of which usually goes into the colloidal form. It is believed that the greater part of the iron transported by rivers is carried in the form of ferric oxide hydrosol, stabilised by colloidal organic substances and adsorbed on clay minerals. These colloids carry positive charges and may be transported long distances without suffering precipitation, provided that the concentration of electrolytes is low, and that negatively charged colloids are not present in sufficient quantities to cause coprecipitation. On entering the sea, such suspensions are liable to become flocculated and deposition soon follows, sometimes in the form of **chamosite** ($3FeO.Al_2O_3.2SiO_2.H_2O$).

In fresh-water lakes and swamps, where stabilising organic colloids are abundant and the concentration of electrolytes is too small to be effective in precipitation of colloidal solutions, deposition of iron appears to be brought about principally by the activities of bacteria and plants. The ferric oxide hydrosols are reduced to the ferrous state and siderite is precipitated.

The accumulation of iron minerals to form an ore deposit requires more than chemical conditions appropriate for large-scale iron precipitation. A marked reduction in clastic input from land is needed, and a mild decrease in current activity to prevent excessive dispersal of the iron minerals. On the other hand, a certain amount of winnowing can concentrate particles, such as ferruginous ooliths, and remove non-ferruginous clastics. Iron ores commonly show a wide range of structures indicative of active currents, such as cross-bedding and ripples.

MARINE IRONSTONES

These deposits include Phanerozoic 'minette-type' chamosite, limonite and siderite ironstones, haematite ironstones and the predominantly Precambrian siderite, haematite and greenalite iron-formations. The Phanerozoic deposits are usually oolitic and variably admixed with other allochems and matrix. A suitable descriptive nomenclature is to deal with the ooliths and allochems independently of the matrix (Fig. 11.2). For an oolitic textured rock with ooliths of chamosite set in a matrix of limonite and siderite the name limonitic sideritic chamosite oolite is appropriate. A shelly calcitic haematite oolite is an ironstone formed of shell débris and haematite ooliths cemented by calcite, and so on. In the absence or near absence of ooliths, the general term mudstone is used with qualifying nouns, such as siderite or chamosite, e.g. siderite mudstone. Mudstones are made up of silt and clay grade particles, the former often being dominant. Limestones often show partial replacement of calcite by siderite, in which case the name sideritic limestone is used. The Liassic Marlstone ironstone of central England is a sideritic chamositic limestone.

Minette-type ironstones

These occur in beds up to ten metres thick in fossiliferous and bioturbated

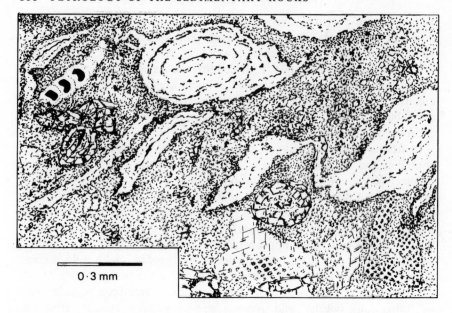

0·3 mm

Figure 11.2 Shelly chamositic sideritic chamosite oolite, Lias, north-west Scotland. Chamosite ooliths, some of which are highly distorted (spastolithic) due to early compaction, and marine shell débris, set in a dense matrix of partly limonitised chamosite and rhombic siderite matrix. Polarised light.

marine successions, about sixty metres thick, the areas of accumulation rarely extending more than 160 kilometres in maximum dimension. Good examples occur within the Jurassic successions of north-west Europe.

The petrology of the ores is comparable in its complexity to that of limestones. Chamosite (an iron-rich chlorite) usually occurs as ooliths, thin coatings around calcareous allochems, and as matrix. It is believed to be chemically precipitated within the top layers of sediment on the sea floor.

Until recently chamosite was not thought to be common in modern marine sediment, but now it is known to occur abundantly at depths down to 150 metres in warm current areas, within ten degrees of the equator. The mineral has been recorded in the Orinoco and Niger off-delta areas and in the Malacca Straits, where it is found as pellets, infillings of skeletal cavities, and replacing the calcite of shell fragments. Transitions from chlorite and biotite into chamosite are known and there is also a tendency for the mineral to alter peripherally into limonite. Limonitisation probably ensues as a consequence of oxidation during reworking of the chamosite grains.

Ooliths in fresh ores sometimes show alternate zones of chamosite and limonite, suggesting a periodicity in reducing and oxidising conditions (Fig. 11.3). Chamosite also readily converts into limonite during subaerial weather-

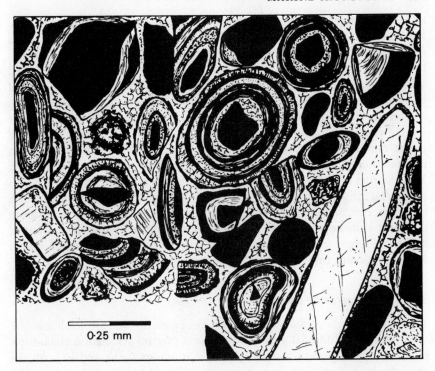

Figure 11.3 Sideritic chamosite oolite, Middle Jurassic, central England. Partially limonitised chamosite ooliths and calcareous shell débris set in a sideritic matrix. Some ooliths are mildly spastolithic and others are much abraded. The margin of the shell fragment to the right is partly replaced by siderite. Secondary limonite is shown black. Polarised light.

ing. In the Cleveland Ironstones (Middle Lias) of north-east England chamosite ooliths often have a white appearance due to decomposition into a mixture of finely divided opaline silica and kaolinite. In some instances the ooliths have totally converted into kaolinite with loss in internal concentric structure.

The diagenetic replacement of chamosite by authigenic pyrite and magnetite is widespread, though not fully understood. Pyrite probably forms at an early stage under reducing, slightly acid conditions prevailing a few centimetres below the depositional interface. Magnetite may prefer more alkaline conditions.

The evidence for primary siderite precipitation is equivocal despite the fact that many fresh mudstones are formed almost entirely of small rhombic crystals of the mineral. Primary siderite ooliths are unknown. Seams in the Cleveland Ironstones contain chamosite ooliths with an external coat of siderite rhombs which penetrate and replace the structure of the ooliths; in extreme examples, all trace of the original chamosite has disappeared. The Northampton Sand Ironstone (Middle Jurassic) of central England contains chamosite ooliths which have been partly sideritised prior to final cementation, and others sideritised after cementation. Calcitic ooliths and shell débris, even detrital quartz and feldspar grains, are also completely replaced by siderite.

Overall, the evidence suggests that many, if not most, minette-type ores are the product of chamosite and limonite precipitation at and very close to the depositional surface, with synchronous and pene-contemporaneous precipitation of diagenetic siderite up to several metres below that surface. The scouring capacity of the bottom currents was adequate, at intervals, to release sideritised fragments and make them available for reincorporation into new, occasionally cross-bedded, layers of sediment. Under these more turbulent water conditions the fragments were cemented, in certain instances, by coarse sparry calcite; some of this calcite is likely to be neomorphic. Redistribution of calcite probably occurs during weathering.

Sedimentary haematite ironstones

These can be considered as a special variety of minette-ore in which haematite is a prominent additional member of the iron mineral association chamosite-siderite-pyrite. The beds usually carry a rich marine fauna and bear all the characteristics of shallow water deposition. The best examples are found in Palaeozoic successions and it is possible that the age is significant, in that deep burial or mild metamorphism might have converted original siderite and limonite into haematite.

The Wabana ironstones (Ordovician) of Newfoundland consist predominantly of sideritic chamositic haematite chamosite oolites. The ooliths are built up of concentric layers of chamosite and haematite, suggestive of alternating reducing and oxidising conditions. The ooliths frequently have a clastic quartz nucleus and, indeed, the ore bodies as a whole belong to an arenaceous facies, limestones being absent. The interbedded sediments are principally sandstones with a chamosite cement.

In contrast, the Silurian Clinton ironstones of the eastern United States and certain Lower Carboniferous ores of Britain are essentially marine limestones in which various constituents have been replaced by diagenetic haematite (Fig. 11.4). The Clinton deposits are mainly oolitic and shelly limestones in which the ooliths and shell fragments are variably replaced by haematite. In these, and similar deposits, there is usually abundant evidence for iron impregnation occurring at or before the time of final deposition. Altered particles are enclosed in clear sparry calcite and dolomite, or reworked haematised fragments are incorporated into unaffected sediment resting directly above. However, whether the iron was precipitated as haematite or some other iron mineral remains an unresolved problem.

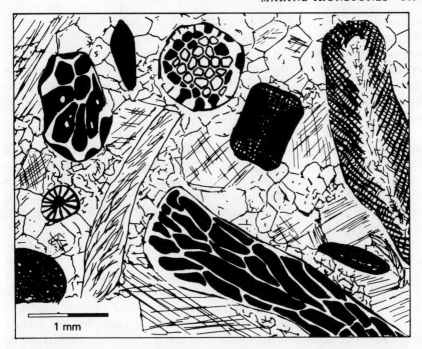

Figure 11.4 Rhiwbina Ore, Lower Carboniferous, South Wales. Bryozoan and crinoid débris impregnated with haematite (black). The matrix, which is virtually devoid of haematite, consists mainly of sparry calcite and small rhombs of dolomite. The ore is no longer mined. Polarised light.

Iron-formations

These are massive units, 30–800 metres thick and traceable laterally for many hundreds of kilometres. The bulk are Precambrian, such as the Biwabik Formation of the Lake Superior region, and formed about 2000 million years ago, though examples are known from younger marine successions in Brazil, Nepal, North America and Europe. It is generally assumed that all are marine in origin.

The beds show evidence of shallow water genesis, such as cross-bedding, ripples, cut-and-fills, sun-cracks and intraformational brecciation, but the characteristic feature is a laminated or banded structure. The banding usually consists of layers, up to two centimetres thick, of chert alternating with iron-rich layers formed of a variety of iron minerals, including magnetite, haematite, siderite, pyrite and greenalite (a pale green ferrous silicate). The banding is not so pronounced in beds containing ooliths and pisoliths. The ooliths exhibit iron and chert zones, whereas the pisoliths consist of an irregular aggregation of greenalite, other iron minerals and chert.

As most iron-formations are Precambrian they have been subject to prolonged diagenesis and, almost certainly, to metamorphic changes, so that the present mineralogy may only be a pale reflection of the sediment as laid down. Siderite generally replaces iron oxides and silicates and, in turn, is replaced by pyrite. Haematite ooliths are replaced by magnetite and vice versa. Greenalite converts into a talc-like ferrous silicate, minnesotaite. It has been claimed that all the ferruginous bands were originally deposited as siderite, though this suggestion is improbable.

It has also been contended that the chert is secondary and formed by the alteration of interbedded shales or quartz sand layers, though traces of clastic particles are now lacking. Many younger minette-ores show an alternation of quartz-rich and iron-rich bands which might produce, on metamorphism, a rock closely resembling a banded iron-formation. The textural similarity between iron-formations and shallow water lime deposits is also remarkable.

In contrast to the views favouring secondary origins for the ores, are those advocating direct chemical precipitation of silica and iron. It is assumed that the bands fairly closely resemble the original sediment in composition. If this is so, the banding might reflect a critical chemical situation whereby the waters in shallow protected areas were near to saturation with respect to silica. During strong evaporative periods, climatically controlled, the balance might have tipped towards silica saturation and precipitation, whereas during weak evaporative periods the conditions might have been appropriate for iron precipitation within the bottom sediments. One of the bases for the chemical argument is that, in Precambrian times in particular, there would be little biological extraction of silica from the waters, thus allowing the silica concentration to increase.

As in minette-ores, it is probably unnecessary to stipulate special conditions governing the supply of iron, though some believe that additional accessions of iron from laterite-mantled continents and submarine thermal sources may have played an important role.

NON-MARINE IRONSTONES

Sediments consisting principally of siderite occur as thin bands or as nodules in certain argillaceous rocks. Some of these sediments, such as the clay ironstones (sideritic mudstones) within the alluvial and deltaic Carboniferous and Mesozoic successions of north-west Europe, are fresh or brackish water in origin. The siderite appears to have been precipitated as a relatively minor constituent of the mud and later, during early diagenesis, to have undergone segregation into nodules and layers.

The clay ironstones of the Coal Measures of north-west Europe are well known. They occur as bands of nodules or as thin continuous beds, buff-brown to gray-brown in colour and compact in texture. Fossils, either plants

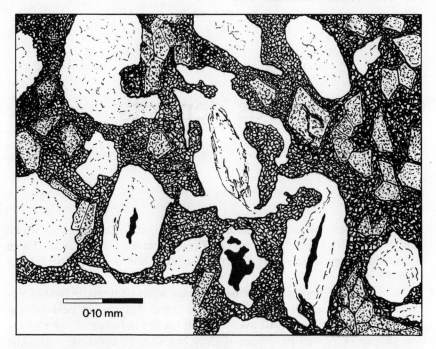

0·10 mm

Figure 11.5 Siderite kaolinite mudstone, Coal Measures, north England. Grossly spastolithic kaolinite ooliths and globules (white) set in a dense matrix of partly limonitised rhombic siderite. There appear to be two generations of siderite, first and most prominent, the fine-grained variety, and second, the larger euhedral rhombs. Polarised light.

or fresh-water bivalves, are common in them, and often the nodule has clearly been segregated round a single organism.

The siderite is fine grained, the usual grain size being about 0·01 milli-metres. Calcium and magnesium carbonates, and a little manganese carbonate, may form about 10 per cent of these rocks, and probably exist in solid solution in the siderite, except in those cases where calcite occurs in the form of fossil shells. In clay ironstones associated with marine beds, pyrite and calcium phosphate become important constituents. Kaolinite (possibly replacing chamosite) ooliths are known at some levels (Fig. 11.5).

12

Carbonaceous deposits

The carbonaceous sediments include coals, peats, and other deposits which consist of the altered remains of plants. These materials can readily be shown to contain recognisable remains of plant tissues, and are demonstrably of vegetable origin.

Coals and peats appear to have been formed almost entirely in fresh-water environments, though it may be borne in mind that certain peats at the present day accumulate in salt marshes, and the groundwater of some of the coal swamps may well have been slightly brackish. In contrast, sapropelic deposits, which are richer in fatty and proteinaceous substances than peats, seem to have been formed under conditions much more diverse than those required for the accumulation of ordinary coal; a few sapropels appear to be thoroughly marine deposits. Peat deposits are accumulating at the present day in environments where oxidation of vegetable matter is retarded either by low temperatures, which lower the rate of bacterial and fungal decay, or by the accumulation of substances which act as antiseptics and inhibit the action of most fungi and bacteria. Oxidation is a surface effect and as sediment accumulates reducing conditions are created at shallow depths in which attack on plant residues is continued by actinomycetae (fungi) and anaerobic bacteria. This microbial activity, in what is essentially a biochemical phase of plant degradation, decreases with increasing depth.

In poorly aerated lakes and ponds black organic mud termed **gyttja** is deposited, the plant matter being recognisably organised.

CLASSIFICATION OF COALS

Humic coals

This is a large group of coaly deposits in which wood is an important constituent; it includes the woody peats, lignites, ordinary bituminous coals and the anthracites. These deposits all contain a similar range of plant tissues. Accumulation of the original plant débris occurred in alluvial and deltaic

Table 12.1   Chemical composition of representative peats and coals.

	Age	Carbon	Hydrogen	Oxygen and nitrogen	Ash
Forest peat	Recent	51·47	5·96	32·68	9·67
Moor peat	Recent	53·59	6·33	27·84	12·24
Lignite (Germany)	Tertiary	57·28	6·03	36·16	0·59
Brown coal (Germany)	Tertiary	61·20	5·17	21·28	12·35
Lias coal (Hungary)	Jurassic	78·08	3·91	7·32	10·69
Cannel (Lancashire)	Carboniferous	80·07	5·53	10·20	2·70
Bituminous humic coal (north-east England)	Carboniferous	83·47	6·68	9·59	0·20
Anthracite (South Wales)	Carboniferous	91·44	3·36	·2·79	1·52

plain situations, apparently under warm, humid conditions. The deposits usually rest directly on seatearths (underclays, fireclays and ganisters).

These are typically banded coals, and have a tendency to develop a small-scale jointing or **cleat**. If we arrange the humic coals in a series beginning with peat, and continuing through lignite, sub-bituminous and bituminous coals to anthracite we find that several progressive changes can be traced through the series. In general, the moisture content decreases, as also do the percentages of oxygen and volatile constituents; at the same time, the percentage both of fixed carbon and total carbon increase steadily (Table 12.1). On this basis, the coals may be divided into a series of ranks, as follows:

	Anthracite
	Semi-anthracite
Increase	Semi-bituminous coal
in	Bituminous coal
rank	Sub-bituminous coal
upwards	Lignite
	Peat

Thus coals of similar chemical composition and of approximately similar behaviour as fuels are grouped together in one rank, and the coals at the top of this list are said to belong to a higher rank than those below.

As a very broad generalisation, one may say that the older the coals are the higher their rank. Carboniferous coals are anthracitic and bituminous, whereas coals of late Tertiary and Recent age are nearly all lignites and peats.

Many geologists maintain that if seams in a single coalfield are compared, the more deeply buried coals are likely to be of higher rank than those nearer the surface. This is often the case.

Probably the most significant feature of rank variation in coals can be determined if a single seam can be traced for a long distance. It is then found that the coal is of higher rank where it has been involved in crumpling, folding and faulting. The rank of coal increases with the intensity of orogenic forces and the natural heat treatment to which the coal has been subjected. By natural heat is meant frictional heat generated during orogenic folding, not magmatic heat.

Regional variations in rank are well illustrated in the South Wales and Pembrokeshire Carboniferous coalfields, where the most highly disturbed seams are anthracites or comminuted anthracites, whereas the less disturbed are bituminous. The variation in rank is particularly noticeable as the anthracitic north-western parts (Ammanford) of the Welsh coalfield is approached.

The Cretaceous and Tertiary coals of the Rocky Mountains field are of bituminous rank where they are involved in the folds of the mountain belt, but where they pass eastwards into the undisturbed region of the prairies, the seams change in character, and are represented by lignites. A similar relationship between orogeny and regional changes in the rank of the coal seams affected can be traced in some of the eastern coalfields. In Pennsylvania, for example, there is a general increase in rank in an easterly direction as the Carboniferous seams become involved in the folds of the Appalachian Mountains.

Sapropelic coals

Amongst the non-woody coals, those types which consist of a macroscopically homogeneous mass of spores, oil-algae, or macerated plant débris, are grouped together as sapropelic coals. The original plant débris accumulated as subaqueous sediment in stagnant pools and lakes, on alluvial flats. Cannel coal and boghead coal, although differing from each other in their vegetable constitution, form natural subdivisions of this group. They are typically unbanded coals which break with a conchoidal fracture and are normally uncleated.

VARIETIES OF COAL

Lignite

This is a low rank coal and is most commonly found in Tertiary and Mesozoic strata. It differs from ordinary bituminous coal in its brown or brownish-black colour, and in its greater content of moisture. Chemically the lignites are characterised by a high oxygen content, and by approximate equality in the proportions of volatile substances and fixed carbon. Because lignites contain from 25 to 45 per cent of water they develop shrinkage cracks and crumble to small fragments when exposed to the atmosphere. Disintegrated or powdered lignite has a strong tendency to take fire by spontaneous combustion.

The texture of low rank lignite is somewhat reminiscent of dried woody peat; the larger plant-remains, such as wood, bark, and leaves, are still recognisable, and are embedded in a dark brown structureless matrix. There are many varieties, some loose and fibrous, others compact and earthy. Lignites are rare in Britain, the only important example being at Bovey Tracey, in Devonshire. Here thick beds are associated with the kaolinitic ballclays of Oligocene or Miocene age. Great deposits of lignite occur in the Cretaceous and Tertiary strata of North America, and there are important beds in Germany and other parts of Europe referred to as braunkohlen.

Sub-bituminous coals (or 'black lignites')

These are intermediate in character between the lignites and bituminous coals. They are black in colour, and, when freshly mined, have rather the appearance of ordinary bituminous household coals, except in their poor jointing. The high moisture content (over 20 per cent) causes them to crumble on exposure to the air, and the disintegrating coal is liable to spontaneous combustion, especially if much pyrite is present. Volatile matter and fixed carbon are present in almost equal proportions, but the moisture content is lower than that of the lignites. As fuels, these coals ignite easily, burn with a bright smoky flame, and are capable of yielding much gas. Coals of this kind are usually of Mesozoic or Tertiary age, and are well developed in the Rocky Mountain region of North America.

Bituminous coals

These include the ordinary household and coking coals. They are generally well jointed or cleated and the joint faces are often covered with a thin layer of pyrite, ankerite or some other mineral. The coals consist of alternating bright and dull layers, the proportions of which vary in different samples. This laminated or banded structure results mainly from the interbedding of four rock-types (lithotypes), each with a different appearance on the fractured surface of a hand specimen and in reflected light inspection under the microscope (Table 12.2).

Fusain is a charcoal-like substance which consists of the maceral fusinite (carbonised wood). It is responsible for the dirty character of ordinary coal, for it is extremely friable and easily reduced to a fine powder. Under the microscope a fibrous and cellular structure is clearly visible. Fusain is believed to result from the alteration of wood in an oxidising environment and most probably was formed from dead branches and trunks exposed to the air.

Vitrain is bright and jet-like with a high lustre and is extremely brittle, breaking with a conchoidal fracture; the layers are almost invariably lenticular in form. The macerals consist of collinite and telinite and are the remains of woody tissue. The tissues are thoroughly impregnated with colloidal humic matter and are believed to have formed from leaves, stems, bark, twigs and logs which fell into extremely stagnant pools of peaty water, and suffered degeneration under anaerobic conditions.

Table 12.2 Classification of coal lithotypes, maceral-groups and macerals.

Lithotype	Maceral-group	Macerals
		Micrinite
Fusain	Inertinite	Semifusinite Fusinite Sclerotinite
Vitrain	Vitrinite	Collinite Telinite
Durain	Liptinite or exinite	Alginite Sporinite
Clarain		Resinite Cutinite

Note. The classification is modified for the soft brown coals. Instead of vitrinite the broadly equivalent term huminite is often used.

Durain forms layers which break with an irregular and rough surface, which is characteristically dull or matt, as distinct from the glassy surface of vitrain. These layers consist of the smaller and more resistant débris of plants, such as macerated fragments of cuticle (cutinite), spore cases (sporinite), isolated bits of resin (resinite) and, in some speciments, remains of algal colonies (Fig. 12.1). Macerals such as fusinite and collinite are also present.

Clarain layers are distinguished by a satin-like lustre, and consist of very thin alternating bright and dull laminae. Microscopic examination shows that the bright laminae consist of thin vitrinite sheets, which are set in a very fine textured groundmass of composite nature. The combination of vitrain with this fine groundmass is known as clarain.

The groundmasses of clarain and durain appear to result from the decomposition of plant débris to a stage when the woody tissues were completely broken down, and only the more resistant materials such as spore cases, cuticles, and resins remained. Bacterial decomposition probably played an important role in this breakdown. This macerated débris is now bound together in a colloidal matrix, formed from the decomposition products of wood and soft tissues.

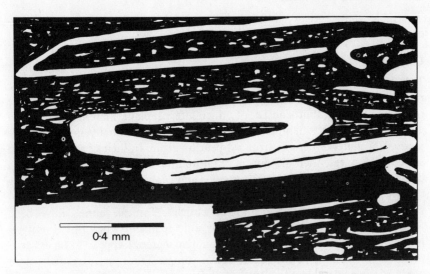

Figure 12.1 Durain. Dull and commonly densely black under reflected and transmitted light. The large white structures are distorted megaspores. Polarised light.

Coals of bituminous rank show a considerable range in oxygen content, though the proportion of hydrogen is more or less constant, varying little from 5 per cent. The proportions of volatile constituents and fixed carbon vary reciprocally within a fairly wide range; coals with much oxygen have a low proportion of fixed carbon (low rank bituminous coal), and those with least oxygen usually have a high proportion of fixed carbon (high rank bituminous coals).

Semi-bituminous coal (or high rank steam coal)

This has a higher heating value than any other kind of coal, and burns with an almost smokeless flame if properly fired. It is a hard, brittle coal with a characteristic prismatic fracture, distinct from the rectangular cleat of bituminous coal. The content of volatile matter is low, being only one-third to one-sixth of the fixed carbon.

Anthracite

This differs from other coals in its extremely high content of fixed carbon, with a correspondingly low proportion of volatile matter, accompanied by low percentages of oxygen and hydrogen. It will ignite only at a high temperature, and burns with a short smokeless flame. Most varieties are clean to handle, owing to the absence of friable fusain, and broken surfaces show a submetallic lustre and conchoidal fracture. Most anthracites have a banded structure similar to that of bituminous household coals, but this is not an invariable characteristic, and massive varieties of anthracite are not un-common.

Cannel coal

This coal is unlaminated, and breaks with a glassy conchoidal fracture rather like that of pitch. Typical examples show little or no recognisable remains of wood, but consist of much-altered plant material containing variable, often high, quantities of miospores, and occasionally an appreciable amount of material derived from oil-bearing algae. In their composition and structure they resemble the durain of bituminous coals, and there is no sharp distinction between dull bituminous coals and cannels, the name cannel being usually applied where woody layers are absent and the bulk of the coal consists of spore material, resin, and cuticles. Typical cannels have been water transported and deposited as organic sediments, but this is not an invariable rule, and many deposits of similar material appear to have accu-mulated in stagnant ponds in the coal swamps. Transported cannels often contain a considerable admixture of clastic sediment, shown in analyses by a high proportion of ash, and such deposits pass into coaly or carbonaceous shales. Cannel coals contain large proportions of volatile constituents, and burn with a bright, smoky flame like that of a candle (Table 12.1).

Boghead coal or torbanite

This is closely allied to some varieties of oil shale and consists essentially of oil-bearing algae mixed with small quantities of detrital sediment. Thin-sections reveal the presence of innumerable translucent yellow bodies, which possibly represent the remains of colonies of an alga resembling the modern *Botryococcus braunii;* these are set in a dark, almost opaque groundmass.

The environments in which boghead coals formed seem either to have been the central parts of large basins in which transport of organic material was restricted or small shallow basins with little input of surface waters. In each case the water was probably relatively free of colloidal matter and well oxygenated.

Jet

This is a compact, coaly substance found in isolated masses in bituminous shales. It always possesses a woody, vitrain-like structure, which may be revealed on etched surfaces or in suitably prepared thin-sections. Most masses of jet appear to have been formed from individual logs of driftwood, which have suffered considerably compression since they were entombed in the sediment, so that the woody tissues, as seen under the microscope, are crushed and distorted. Occasionally a silicified core is found, preserving the woody tissues in their original, uncrushed form. Like vitrain, jet is believed to be derived from wood which at an early stage became waterlogged in an entirely anaerobic environment.

OIL SHALES

There is no completely satisfactory definition of the name oil shale. Perhaps the most suitable is 'a fine-textured rock, normally laminated, containing organic matter from which substantial amounts of oil can be extracted by heating'. Free-flowing oil is unusual though frequently there are small veins, blebs and pockets of solid and viscous bituminous matter, such as asphalt, gilsonite and ozokerite, which have become emplaced secondarily. Carbon disulphide dissolves this bituminous matter but the bulk of the organic content in the shales, commonly referred to as **kerogen**, remains unaffected.

Oil yields ranging from 38 litres per tonne to 550 litres per tonne have been recorded from a range of rocks – limestones, marlstones, shales, silt-stones and impure coals – all collectively referred to as oil shales. The Scottish Lower Carboniferous oil shales, which are predominantly true shales, yield on average 90 litres per tonne; whereas the Eocene Green River Formation oil shales of Colorado, Wyoming and Utah, mainly marlstones and dolomitic limestones, yield on average 125 litres per tonne.

Individual seams of oil shale vary in thickness from a few centimetres to three metres and sometimes extend in uniformly bedded fashion for con-siderable distances. Autochthonous coal seams are seldom interbedded.

Kerogen

The term kerogen ('oil-former') was first used with reference to the car-bonaceous matter in Scottish shales which gave rise to crude oil on distillation. The evidence favouring animal matter as a significant contributor is slim, and most authorities claim plants as being the dominant source. In some cases the principal plant constituents are spores, e.g. Permo-Carboniferous shale of Tasmania, in others algae, e.g. Ordovician shale (kukersite) of Estonia and Eocene Green River formation of North America. But, in most deposits the yellow, orange, red and brown kerogenous matter is not recog-nisably organised, and is very difficult to identify.

Chemically, kerogen consists of a mixture of large organic hydrocarbon molecules, such as those constituting certain waxes and fats (lipids), certain oils and pigments (isoprenoids and terpenoids) and certain resins (steroids).

Environments of deposition

Large lakes These are usually extensive land-enclosed basinal areas being particularly well developed where there has been block faulting and tectonic warping. The best known examples are the Eocene Lakes Uinta and Gosiute of Colorado, Utah and Wyoming, the early Carboniferous Lake Cadell of the eastern end of the Midland Valley of Scotland, and the lakes of New Brunswick, Canada, in which the Carboniferous Albert Shale was deposited.

Tuffs and other volcanic rocks are sometimes dispersed among the oil shales many of which are carbonate-rich or closely interbedded with fine-grained limestones. Extensive evaporite beds may form during prolonged phases of aridity and desiccation.

Shallow, marine seas These are of continental platform or shelf depths. The oil shales deposited within them are characteristically siliciclastic, poor in oil yield, and extend over several hundreds or thousands of square kilometres. The oil shales are often associated with limestones, cherts, sandstones and phosphatic deposits, though this association can also be found in large lakes.

Small lakes Small depressions in swamps have plant and inorganic detritus swept into them which is admixed with indigenous organic materials. If conditions are appropriate canneloid or boghead type oil shales may form. Certain of these may be thick and of high grade, as in the Tertiary coal succession of Fushun, Manchuria.

Scottish oil shales

Important beds of siliciclastic oil shale are contained in the Lower Carboniferous Oil-Shale Group (upper part of the Calciferous Sandstone Measures) of the Lothians and Fifeshire (Table 12.3). Typical specimens are fine-grained

Table 12.3 Chemistry of oil shales.

		Scottish shale (siliciclastic)	Green River shale, Colorado (carbonate-rich)
	Ash (%)	78	65
	SiO$_2$	56·7	42·4
	Al$_2$O$_3$	25·0	10·5
Chemical	Fe$_2$O$_3$	9·9	4·7
composition	CaO	2·7	23·5
of ash (%)	MgO	3·1	9·3
	Na$_2$O + K$_2$O	2·0	7·6
	SO$_3$	0·9	2·0

rocks, brown or black in colour, and showing a brown streak when scratched. The shale is finely laminated and commonly varved, but this structure is not always visible in hand specimens. Beds having the appearance of being internally crumpled and slickensided are known by the miners as 'curly shale', as distinct from the normal undistorted 'plain shale'.

Under the microscope the rock is seen to contain algal material of planktonic habit, a high proportion of fungal- and bacterial-remains related to this material and a small proportion of animal-remains. This autochthonous material is intermingled with a variable amount of detrital débris including partly disintegrated wood fragments, miospores of land plants and resinous and other organic matter probably precipitated from colloidal suspension. The presence of so much well preserved cellular and structural organic matter indicates accumulation under stagnant bottom conditions which eventually arrested further decomposition.

The oil shales accumulated in Lake Cadell, which was land-locked for considerable lengths of time and occupied, at its known maximum extent,

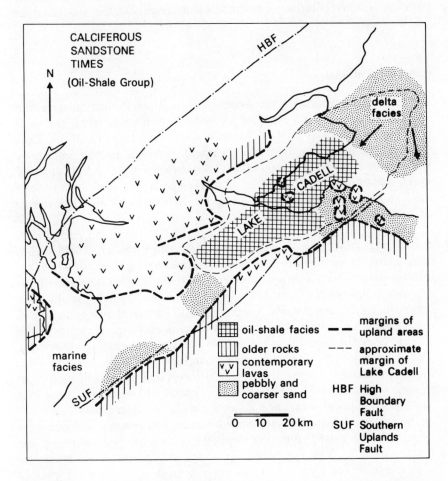

Figure 12.2 Palaeogeography of Scottish Oil-Shale Group times. Although the bulk of the clastic material was fed into the lake via the major deltas to the northeast, there was a steady, but restricted, input from the uplands bounding the lake.

some 3500 square kilometres of the Edinburgh Basin, a fault-bounded sub-siding area (Fig. 12.2). The fresh-water lake existed as a major physiographic unit on at least nine occasions, and on each occasion anything up to five separate oil shale seams, now 0·2 to 6 metres thick, were formed. Con-temporary vulcanicity frequently modified local depths and the shape of the lake.

There is considerable evidence for repeated partial desiccation of the lake. Sun-cracking and brecciation of some of the oil shales and associated algal and oolitic micrites, and siltstones is common. Moreover, many of the limestones show early diagenetic dolomitisation. The constant, though irregular, influx of river water from a major delta complex to the east appears, however, to have effectively prevented complete desiccation and the mass precipitation of salts.

Green River oil shales

The sediments constituting the Eocene Green River Formation were laid down in tectonic basins occupied by several continental lakes, the most important being Lake Uinta (Colorado and Utah) (Fig. 12.3) and Lake Gosiute (Wyoming). At its maximum extent in middle Eocene times Lake Uinta covered an area of 54000 square kilometres (cf. Lake Gosiute at 32000 square kilometres) this interval being marked by a very distinctive zone of rich and thick oil shales yielding oil up to 350 litres per tonne. The zone is known as the 'Mahogany ledge'.

There are two distinct types of oil shale in the Formation and their charac-teristics may have been determined by the depth of water in which they accumulated. Type 1 is least common, black weathering to bluish-white, has organic matter which is recognisably organised and is often sun-cracked and brecciated. It probably formed in very shallow water. Type 2, the abundant variety, is light to dark brown weathering buff, carbonate-rich and contains completely disorganised yellow and brown translucent organic matter (Table 12.3). It formed in waters of moderate depth, probably between 10 and 20 metres. The yellow material occurs in long thin bands parallel to the laminae, while the brown material occurs in irregular stringers and masses. In both types of shale the recognisable organic remains appear to be planktonic algae with low contributions of spores and pollen grains from higher plants.

The carbonate-rich oil shales are lithologically marlstones and dolomitic limestones and usually show fine varve-like laminations with sharp top and bottom contacts. The varves average a few millimetres in thickness and comprise a graded layer which is coarser-grained and organically poor at the base and finer-grained, organically-rich at the top. The varving is explained by differential gravity settlement of mineral and organic constituents from seasonally and chemically stratified lake waters, the production peaks of the

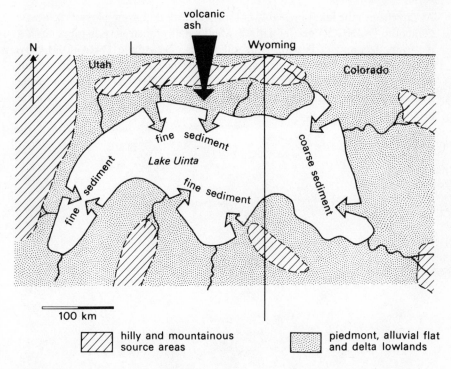

Figure 12.3 Palaeogeography of Green River formation times, Utah and Colorado. The approximate position of the shoreline during Middle Eocene times is indicated. The oil shales formed mainly during the phases of lake expansion. Evaporites were precipitated at several central sites during phases of contraction.

heavier specific gravity clastic and chemical carbonates and the lower specific gravity plankton being in the summer season. The total evidence indicates relatively long warm summers and cool moist winters, the mean annual temperature being estimated at 20 °C.

As with the peak phases of Scottish oil shale deposition, it is possible to trace the change of facies from the deepest parts of the lakes (oil shales) towards the shore (algal, oolitic and ostracod limestones, shell marls) and eventually into the marginal swamps and deltas (clays, siltstones, sandstones, palaeosols and coals). At certain times, tuffs and tuffaceous sediments were deposited along the lake margins, and these materials may have played some part in the precipitation of thin seams consisting almost entirely of euhedral analcite (a sodium zeolite). Magadi-type chert (p. 173) also occurs as continuous thin beds interlayered with the oil shales.

The desiccation stages in the Green River lakes were characterised by extensive precipitation of a host of unusual salts at their depositional centres.

At these times the lakes probably comprised several discrete playa-type areas. Nahcolite ($NaHCO_3$) occurs as individual small grains, masses more than a metre in size, and beds within the oil shales. Dawsonite ($NaAl(OH)_3CO_3$) is usually disseminated as small grains in the shales. In Lake Gosiute, about 90 000 million tonnes of bedded trona ($Na_2CO_3.NaHCO_3.2H_2O$) or trona interlayered with halite, with almost an equivalent amount of shortite ($Na_2CO_3.2CaCO_3$) were deposited.

13

Phosphatic deposits

Phosphates are present in small quantities in the soil, and in the waters of rivers, lakes, and the sea. The bulk of this material is kept in circulation, being taken up in small amounts by plants, and in much greater quantities by animals. In the normal course of the phosphorus cycle, a considerable proportion is returned to an inorganic condition through the decomposition of excrement, or upon the death of the animal; the cycle then begins afresh. The ultimate source of the phosphorus available to organisms is doubtless to be found in the small crystals of apatite which are widely distributed in the crystalline rocks, and which release a constant supply in the course of weathering. At the other extreme, phosphorus is constantly being removed from circulation by the burial of animal débris to form phosphatic sediments.

The mineralogy of the sedimentary phosphates presents considerable difficulties, principally because these substances are rarely crystallised, and most deposits contain impure mixtures of amorphous phosphate with detrital and calcareous matter. The larger marine deposits, i.e. phosphorites, consist principally of calcium phosphates, the exact compounds present in some cases not being clearly identifiable. Some phosphate rocks approach the composition of collophanite, the calcium triphosphate, $Ca_3(PO_4)_2.H_2O$; others consistently show a minor content of fluorine, thus approaching the composition of fluor-apatite, $Ca_5(PO_4)_3F$. Combined calcium carbonate appears to be present in some phosphorites, probably as the mineral dahllite, $4Ca_3(PO_4)_2.2CaCO_3.H_2O$. Dahllite may contain up to 1 per cent fluorine. Carbonate-hydroxyapatites with more than 1 per cent fluorine are referred to as francolite. Guano contains a large number of distinct compounds; many of these are phosphates of calcium, magnesium, and ammonium. Several other terrestrial phosphate deposits, formed by the diagenesis of aluminous materials such as clay, bauxite, or the alumino-silicates of igneous rocks, give analyses approaching certain aluminium phosphates, such as variscite, $AlPO_4.2H_2O$. In fresh-water environments, vivianite, $Fe_3(PO_4)_2.8H_2O$, is frequently found in deposits of bog-limonite, or in associated peats or clays.

In a few types of deposit, such as bone-beds and recent accumulations of guano, the original form of the phosphate is readily apparent, but in the majority of occurrences the phosphates have been redistributed and owe their present form to diagenetic processes. The common phosphate minerals are very difficult to precipitate directly from sea-water.

Phosphatic nodules

The phosphate of newly deposited sediments shows a strong tendency to migrate interstitially during diagenesis, and to become concentrated in the form of nodules. In most of these nodules the phosphate appears to have permeated the original sediment, so that the non-phosphatic part of the nodules consists of clay, lime-mud or glauconite-bearing sand resembling that of the enclosing rock. Such nodules have an earthy texture and are usually gray, brown or black in colour.

Phosphate nodules and slabs, up to a metre long and weighing up to seventy kilograms, are present on submarine ridges and basin slopes off the Californian coast. This and similar areas throughout the world, such as the west coasts of southern Africa and South America, are typified by a slow rate of deposition and oxidising conditions. Moreover, they are areas where cool water currents (about 12–16 °C), rich in phosphate ions, rise from the oceanic depths, spread over the adjacent shelves and mix with warmer surface waters (about 15–22 °C). Nearly all the deposits occur at depths between 30 and 300 metres, where they are associated with manganese oxides and glauconite. Many of the phosphate minerals have the properties of collophanite, but there are significant amounts of other phosphate minerals, such as francolite and dahllite.

Although the present oceanographic conditions are suitable for nodule formation, evidence has accumulated that many originated by phosphatisation of sea bottom sediment during Eocene, Plio-Miocene and Pleistocene episodes of marine regression. The distribution of nodules on the present ocean floor reflects, therefore, both modern and ancient events. Some of the ancient nodules may have been redistributed during subsequent sea level changes.

Diagenetic phosphatic nodule beds are common in Phanerozoic successions, good examples occurring in Upper Cretaceous beds in north-west Europe. The basement marl beds and hardgrounds of the Chalk often contain nodules which have replaced carbonate and have an internal zoned structure, in various shades of brown. This structure suggests several phases of phosphatisation alternating with periods of non-precipitation, the latter being caused by partial exposure on the sea bottom.

Bedded phosphorites

Many Phanerozoic phosphatic deposits are of local character or form special phases within formations of a different nature, such as limestones or shales.

In a few regions, however, phosphatic deposits are developed on a much vaster scale, and constitute independent marine formations covering a considerable area. These are called phosphorites and contain more than 18 per cent P_2O_5. One such, of Upper Cretaceous–Eocene age, extends across North Africa from Mauritania to Syria and appears to have formed at the edge of a shelf sea bounding a Tethyan deeper water region to the north, approximately on the site of the present Mediterranean Sea. A similar large occurrence is to be found in the Rocky Mountain region of North America, where the Phosphoria Formation (Permian) maintains its shelf-edge character over an area of several thousand square kilometres. The purer beds contain up to about 80 per cent of phosphate and are interbedded with dark coloured phosphatic shales, bedded cherts and occasional impure limestones. Nearly all the phosphate is in the form of fluor-apatite with a small proportion of calcium carbonate. Many of the phosphorite beds have a peculiar granular structure, which is commonly described as pelletoid or oolitic. This structure has been interpreted either as coprolitic or oolitic in origin, with metasomatism by phosphatic solutions occurring very soon after deposition.

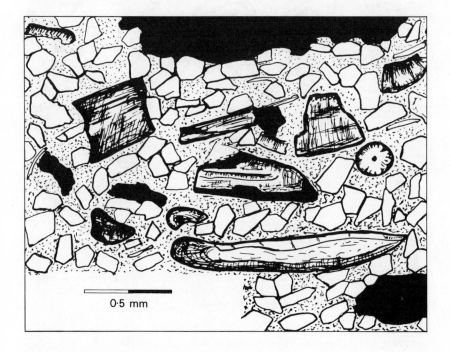

0·5 mm

Figure 13.1 Rhaetic Bone Bed, Gloucester, England. A fine-grained calcareous sandstone containing angular and brown bone fragments preserved in collophanite. Authigenic pyrite patches (black) are common. Polarised light.

Bone beds and fish beds

Deposits containing abundant vertebrate remains are often highly phosphatic, and the recognisable fish scales or bones commonly act as nuclei for the secondary concentration of calcium phosphate to form nodules. Examples of this kind of deposit are to be found locally in the Rhaetic Bone Bed of south-western England and Silurian Ludlow Bone Bed of the Welsh Border-land (Fig. 13.1).

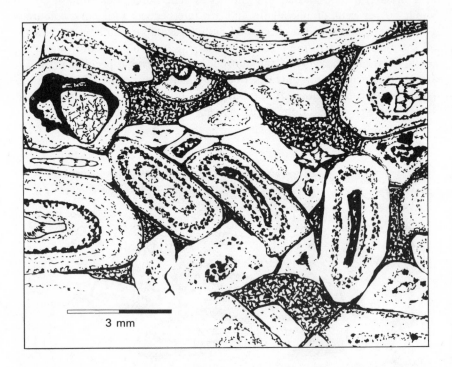

Figure 13.2 Phosphatic pebble bed, Lower Carboniferous, Scotland. Phosphatic pebbles with bone nuclei set in a matrix of granular ferroan dolomite. Polarised light.

Phosphatic pebble beds and pebble phosphates

These beds consist largely of derived material, washed out of nodular phosphate beds, or formed by the erosion of bedded phosphorites (Fig. 13.2).

In some cases, the pebbles have suffered very little abrasion, but frequently they have undergone prolonged rolling, in process of which they have been worn down to spherical or ellipsoidal shapes.

The pebble phosphates of Florida are deposits of this kind which have been concentrated from older sediments; they have been redeposited as river gravels, and are now found on the flood plains of the present streams, and also in the gravels of river terraces. The phosphate pebbles occurring in some Tertiary beds of Maryland, USA, were originally composed of calcite, but were partially replaced by collophanite at an early stage in their diagenesis.

Phosphatic pebble beds containing well rounded fragments of earthy-textured phosphate are familiar features in some developments of Jurassic and Lower Cretaceous strata in Great Britain, and are almost invariably associated with local erosion and non-sequences.

Guano

The principal deposits of guano are formed on rocky islands frequented by sea-birds, the greatest accumulations being found in the dry trade-wind belts, as in the West Indies and in the islands of the eastern Pacific Ocean. When freshly formed, guano is a complex mixture of phosphates, nitrates, and carbonates, with various organic substances in the chemical sense. A slow, diagenetic change takes place after deposition, involving the loss of soluble, volatile, and oxidisable substances, and the concentration of calcium tri-phosphate. Soluble phosphates, leached out in the process, are carried down-wards and progressively replace the underlying foundation of sedimentary or igneous rocks.

14

Evaporites

Under this heading are included deposits whose origin can be referred directly to processes of precipitation and crystallisation from saturated solutions. Although substances of this class often occur as lenticular or irregular masses, the chemical deposits are more usually found regularly interstratified with fragmental or organic sediment, most of the well bedded occurrences having been laid down in lakes or in isolated portions of the sea, while the other types are commonly of terrestrial origin.

The most important of the processes is undoubtedly evaporation. This takes place extensively in salt lakes or in partially isolated marine basins, the products thus formed being referred to as **evaporites** or **salt deposits**. Powerful evaporation combined with a comparatively low rainfall is essential for the accumulation and preservation of such deposits; similar conditions in a large measure also determine the positions of deserts, with the result that salt deposits are commonly associated with fragmental deposits of desert facies.

The most important salt deposits occurring on a large scale in nature are the chlorides and sulphates of sodium, potassium, magnesium, and calcium (Table 14.1). These sometimes form simple salts, such as sodium chloride and calcium sulphate; but they are also found in the form of double salts, or even as more complex combinations. Since the water of the open sea is nowhere saturated with the salts dissolved in it, it is clear that salt deposits cannot be formed in the sea at the present time; and there is no reason to believe that the proportion of salts in the open sea was ever higher in the past than at present.

Partly confined bodies of water may suffer considerable losses by evaporation, especially in areas of high temperature and semi-aridity, with the result that the salts in solution become sufficiently concentrated for crystallisation to begin. Present day sea-water will first precipitate carbonates then, at salt concentrations beyond four times normal, sulphates form, beyond twelve times normal halite, beyond sixty-four times normal magnesium-potassium salts, and beyond one hundred and twenty times normal bischofite.

Many marine evaporite basins show a crudely regular distribution of these

Table 14.1 Principal sedimentary chlorides and sulphates.

Class	State	Species	Formula
		Halite	NaCl
	Anhydrous	Sylvite	KCl
CHLORIDES		Bischofite	$MgCl_2.6H_2O$
	Hydrous	Carnallite	$KMgCl_3.6H_2O$
		Glauberite	$Na_2SO_4.CaSO_4$
	Anhydrous	Anhydrite	$CaSO_4$
		Barytes	$BaSO_4$
		Langbeinite	$K_2SO_4.2MgSO_4$
SULPHATES		Mirabilite	$Na_2SO_4.4H_2O$
		Kieserite	$MgSO_4.H_2O$
		Gypsum	$CaSO_4.2H_2O$
	Hydrous	Polyhalite	$Ca_2K_2Mg(SO_4)_4.2H_2O$
		Hexahydrite	$MgSO_4.6H_2O$
		Epsomite	$MgSO_4.7H_2O$
		Kainite	$4KCl.4MgSO_4.11H_2O$

primary precipitate evaporite facies, both in space and time. Calcium sulphates are precipitated usually towards the edge of the basins and form the lower layers within a given evaporite succession. Chlorides are precipitated towards the central, more depressed parts of the basins and form the upper layers in given successions (Figs 14.5 and 14.6). In the potash-bearing facies the regularity in distribution may be affected by secondary metasomatic changes caused by residual interstitial brines. Further irregularities in the primary pattern of deposition may be incurred by dissolution and replacement of salts during phases of prolonged subaerial exposure between marine inundations. Lowered sea levels could account for this exposure.

SALINE DEPOSITS OF SALT LAKES AND INLAND SEAS

Inland drainage systems are found at the present day in the drier parts of the world where the surface drainage is entirely balanced by evaporation, and there is no excess of water to overflow to the ocean. The salts carried by rivers and springs into such basins gradually accumulate in the lake water until they reach such a concentration that crystallisation begins. The composition of these solutions varies considerably in different basins, both in the nature of the dissolved salts and in their concentration (Table 14.2). In most cases there are marked differences from the mixture of salts present in seawater. It is true that the salt of a few lakes, such as the Great Salt Lake of Utah, shows a curious similarity to that of the ocean, but the waters of most inland lakes are distinguished by individual peculiarities of composition, and

Table 14.2   Salt content of selected lakes (*in parts per 1000*).

	Lake Urmiah, Iran	Great Salt Lake, Utah	Dead Sea	Lake Elton, Russia	Carson Lake, Nevada	Lake Domo-shakovo, Siberia
Sodium chloride	190·47	118·63	63·86	38·3	64·94	3·55
Magnesium chloride	5·22	14·91	163·67	197·5	—	6·08
Calcium sulphate	1·81	0·86	0·78	—	—	2·84
Magnesium sulphate	8·00	—	—	53·2	—	—
Sodium sulphate	—	9·32	—	—	13·76	132·82
Sodium carbonate	—	—	—	—	29·25	0·21
Total salinity	?05·50	143·72	228·31	289·0	107·95	145·50

are often characterised by the predominance of salts or radicals which in sea-water are subordinate or present only in small traces. Alkaline lakes usually contain a considerable proportion of sodium carbonate; other lakes may be dominated by sodium or magnesium sulphate, or may contain unusually large quantities of magnesium chloride. A few lakes are remarkable in containing notable quantities of borates, whilst others show abnormal proportions of bromine, potassium, or dissolved silica.

The analyses in Table 14.2 of the waters of some typical salt lakes illustrate only a part of the wide variations in composition that are possible; these are for the most part cases of extreme concentration, in which the water is saturated, at any rate for some of the salts, and deposition is actually taking place.

These figures indicate the existence of two fairly well marked types, viz.: (1) **salt lakes proper**, with dominant chlorides, of which the Dead Sea is a good example; (2) **bitter lakes**, with sulphates and alkaline carbonates, e.g. Carson Lake, Nevada. Lake Elton is intermediate between these two groups, being rich in both chlorides and sulphates.

The Dead Sea

The Sea was initiated as a marine embayment in Pliocene times along the line of the Great Rift Valley, but soon lost its oceanic connection, so that by Pleistocene times it was totally land-locked. A continental salt sequence, at least 4000 metres thick, was laid down in the subsiding rift, and this style of deposition has persisted through to the present day. Quaternary climatic changes have caused periodic diminished inflow and increased evaporation, which in turn have caused the level of the lake waters to fluctuate. During low inflow phases the salts became increasingly concentrated and, as now, were deposited in greater volumes at certain seasons of the year.

It has been shown that the waters of the river Jordan, which is the principal tributary of the Dead Sea, are unusually rich in dissolved mineral matter, because it flows through a region largely composed of salt-bearing rocks of Pleistocene, Tertiary and possibly Cambrian age. Besides this source of supply there are, in the neighbourhood, many mineral springs, some of which are hot and connected with the volcanism that accompanied the subsidence of the rift zone in which the Dead Sea lies. It is also believed that there are subaqueous mineral springs, some of which yield bituminous products. A remarkable feature of the Dead Sea water is the presence of a large proportion of bromine, and the Cl:Br ratio in the modern halite deposits at the southern end reaches as much as 2500:1. This figure contrasts with halite and allied chloride deposits of modern and ancient marine basin origin, which generally have a ratio falling between 3000 and 15 000:1.

As shown by analysis the waters of the Dead Sea are extraordinarily rich in magnesium chloride. Since this salt is much more soluble than sodium chloride, and the water is not yet saturated with it, compounds of magnesium are not being deposited. On the other hand, since the presence of magnesium chloride diminishes the solubility of sodium chloride, the water is saturated for sodium chloride. Consequently crystals of common salt, together with gypsum, are abundantly present in the muddy deposits now being laid down on the floor of the lake, and the sodium and calcium salts brought in by the Jordan and other rivers are at once precipitated.

The principle here enunciated is very important in the study of salt lakes; and it accounts for the fact that common salt is now being deposited in many lakes in the waters of which very different proportions of this salt are contained, since the saturation point for any one salt is controlled by the proportions of other salts present in the solution.

Playas (salinas)

The deposition of salts in playas, which often form in arid and semi-arid interior drainage basins, follows a comparatively simple pattern and frequently illustrates how the chemical evolution of brines in such an environment is mainly controlled by the bulk composition of the parent water and the degree of evaporation.

A good modern example of an inland salina occurs in the Saline Valley of California where a zonal distribution of evaporites intermingled with the sands and muds can be detected (Fig. 14.1). The evaporite minerals also occur as surface efflorescences. Water flowing into the playa emanates from springs and mountain streams but, by the time it reaches the playa area, is subsurface. Initially, it is relatively rich in sodium, calcium, sulphate and bicarbonate ions but the response to a high rate of evaporation ensures progressive enrichment in sodium, chloride and sulphate ions towards the playa centre. The sequence of events is: first, the precipitation of calcite in alluvial fans adjacent to the playa which decreases the concentration of calcium and bicarbonate ions in the brines; second, the precipitation of

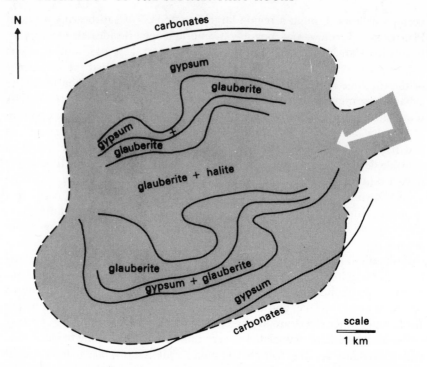

Figure 14.1 Zonal distribution of evaporite minerals in Saline Valley playa lake. The distribution applies to the minerals in the upper four metres but below the efflorescent crust.

gypsum at the playa edge, this modifies the sulphate ion concentration; and finally, the concentration of sodium and chloride ions in the central parts which progressively gives rise to the deposition of glauberite and halite in variable proportions. Contemporary precipitates of analcite ($NaAlSi_2O_6.H_2O$) and sepiolite ($Mg_2Si_3O_8.nH_2O$) are distributed through the playa sediments.

The various salts are usually precipitated from capillary pore waters above the water table and form hard crusts, a metre or so in thickness. Around the margins of the playas even sand dunes may become effectively stabilised by this process. The surface crusts show extensive desiccation cracking and pressure ridging due to the crystallisation of new salts. They also show corrosion into pseudo-ripples, cusps and channels, suggestive of spasmodic free flow of surface water. Salt stalactites and tufa-like cones and terraces are associated features.

The Caspian Sea

Around the Caspian Sea the formation of salt deposits takes place in partially isolated areas having limited communication with the main body of water.

Under such geographical conditions and where high temperature causes strong evaporation, while a small rainfall limits the supply of fresh-water, the concentration of salts may be carried to the point of saturation and precipitation. A classic example is the Karabogaz Gulf, some 18 000 square kilometres in size, on the eastern side of the Caspian Sea. The waters of this Gulf are about three metres deep and have a composition very different from that of the Sea (Table 14.3).

Table 14.3 Salts in the Caspian Sea and in the Karabogaz Gulf.

	Caspian	Karabogaz
Sodium chloride	8·116	83·284
Potassium chloride	0·134	9·956
Magnesium chloride	0·612	129·377
Magnesium sulphate	3·086	61·935
Salinity per mille	11·948	284·552

The Gulf is separated from the Caspian by a narrow strait some ten kilometres long, 100–150 metres wide and about six metres deep. A continual current runs through the strait into the Gulf to supply water lost by evaporation, hence there is a continual addition of salts in solution. Chemical analyses furnish clear evidence that a great concentration of salts of sodium and magnesium has taken place in the waters of the Gulf over the last three or more decades.

Sodium and magnesium salts, such as halite, epsomite and astrakhanite ($MgSO_4.Na_2SO_4.4H_2O$), are now being precipitated over about 75 per cent of the water covered area (about 10 000 square kilometres). Prior to the 1930s the principal mineral deposited was glauberite but, since then, the proportion of other salts has increased due to increase in brine concentration. Halite was first precipitated in 1939.

During present summers halite, epsomite and astrakhanite are deposited, mainly in the northern and eastern parts of the Gulf where there are the highest brine concentrations. In winter, mirabilite with a little epsomite is precipitated. Mirabilite is also deposited in the southern and western parts during the winter months. Gypsum is forming fairly continuously at the margins of the Gulf, particularly on the western side adjacent to the strait. However, the boundaries between the evaporite facies are variable in position, depending on the season, volume of inflow and the displacement of surface brines by the intensity of the prevailing winds.

Beneath the modern salt layers are three older layers separated by carbonate beds. It appears that this triple succession reflects late-Quaternary rises and falls in surface level of the Black Sea, the carbonate oozes being laid down during transgressive, high sea level phases, and the salts being precipitated during regressive, lower sea level phases.

Marine salt pans

Crystallisation of sodium chloride is taking place at the present day in numerous salt pans or marine salinas. The salt ponds of the Bahama Islands are shallow pools lying behind the coast, and in most cases separated from the sea by a pervious barrier. One of the best known examples in the Mediterranean region is the salt lake near Larnaca on the island of Cyprus. Strong evaporation leads to concentration of salt and keeps the level of this lake a metre or so below that of the open sea, from which it is separated by a barrier of unconsolidated sediments. A pervious layer of sand and gravel allows a slow inward seepage of salt water to take place.

Persian Gulf

The high intertidal and supratidal (coastal sabkha) zones of the arid Trucial coast of the Gulf are prograding areas of lime accumulation, several kilometres wide. Dolomitisation causes widespread early diagenetic changes in

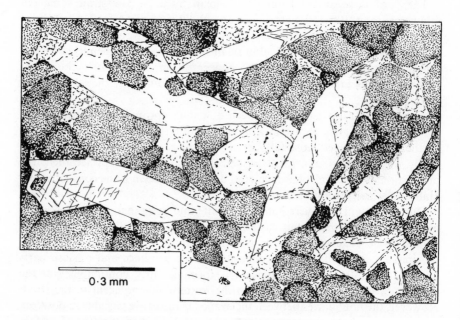

0·3 mm

Figure 14.2 Replacement of authigenic gypsum by calcite. The original rock was a Jurassic pellet limey mud, probably laid down in an intertidal-lagoonal situation. At an early stage in the lithification, gypsum was precipitated from circulating pore waters enclosing some, and displacing other, pellets. At a later diagenetic stage the gypsum was replaced by calcite, some of this extending beyond the confines of the original gypsum grain to replace the original rock (far left). Polarised light. After West (1964).

the sediment, but there are other equally important early changes involving the precipitation of anhydrite, gypsum, halite and celestite from capillary movements of groundwater above the water table. The water is mainly derived from the sea and moves laterally from the intertidal zone into the sabkha zone, where surface evaporation converts the 'normal' salinity pore water into brine.

In the high intertidal flat lime sediments gypsum forms, commonly beneath algal mats (stromatolites). The gypsum and stromatolites persist as layers up to thirty centimetres thick a metre or so below the prograding sabkha surface.

From the more concentrated capillary brines above the water table in the sabkha zone anhydrite is precipitated in preference to gypsum. The anhydrite crystals progressively coalesce into nodules and nodular layers, some of which extend over wide areas of the sabkha. A reticulate 'chicken-wire' structure characterises the layers. During the act of displacing the pre-existing carbonate grains of the lime deposit, the nodules commonly develop a marked intestine-like contortion which is referred to as enterolithic folding.

Both the near-surface anhydrite and lower level gypsum in the sabkha areas are susceptible to further diagenetic changes. Gypsum is known to replace anhydrite if the circulating pore waters become less concentrated with salts, and vice versa. Calcite, dolomite, celestite and chalcedonic silica are known to replace crystals of anhydrite and gypsum (Fig. 14.2). Bladed grains of secondary gypsum, up to twenty centimetres long, and 'desert-rose' aggregates typify the surface of sabkhas. These grains commonly enclose quartz sand.

ANCIENT EVAPORITES

Marine evaporites are present in strata of late Precambrian age and occur throughout the geological column up to the present day. Some of the oldest are in Pakistan, Iran, and Siberia. The United States has evaporites in every geological system from the Ordovician onwards, and at certain times the accumulations were extensive and thick. Over 260 000 square kilometres of the eastern Great Lakes region is underlain by Upper Silurian gypsum, anhydrite and halite beds which, in places, reach thicknesses of 450 metres. Thick anhydrite and gypsum beds are also present in the Carboniferous of Utah (1200 metres), the Permian of south-eastern New Mexico (1350 metres) and the Cretaceous of south-central Florida (1800 metres). In the Whitby district of north-east England at least 480 metres of Permian evaporites have been proved in bore holes (Table 14.4).

In many of these successions there is evidence that some of the deposits were laid down in current-swept, relatively shallow seas and under sabkha-type conditions. Ripple marks partly of corrosive origin, cross-bedded units up to thirty centimetres thick, minor unconformities, desiccation cracks and

Table 14.4 Permian evaporite succession of north-east England.

		Thickness (*m*)
Upper Permian marl		180
Top anhydrite		0·5–1·2
Salt clay		2–4
Upper evaporites	Upper halite	15–27
	Potash zone	0–9
	Lower halite	12–18
	Anhydrite	5–8
	Carbonate	0·5–1·2
Carnallitic marl		9–18
Middle evaporites	Upper halite	0–5
	Potash zone	3–4
	Lower halite	24–83
	Halite-anhydrite \ Anhydrite	15–27
Upper Magnesian Limestone		35–55
Lower evaporites	Upper halite-anhydrite	12–35
	Upper anhydrite	12–201
	Lower halite-anhydrite	14–42
	Lower anhydrite	45–93
Lower Magnesian Limestone with some anhydrite Basal sands, marls and breccias		±110

'shale ball' pebbles, associated with marine fauna and flora, are found towards the edge of basins. In the Upper Permian succession of eastern Yorkshire, England there are repeated developments of lenticular and nodular anhydrite, exhibiting displacement structures and interbedded with fine-grained, algal dolomites. The deposits were laid down at the western margin of the Zechstein Sea in what seem to have been extensive sabkhas and coastal lagoons. Over one period, represented by the diachronous Billingham Anhydrite formation, the sabkha zone alone had a width perhaps as much as eighty kilometres.

A feature of ancient sabkha successions, as in parts of the Lower Carboniferous of the Midlands and Lake District, England, are rhythmic units, about thirty centimetres thick, each constituted of at least a basal sparsely fossiliferous, algal limestone passing upwards into a sulphate-rich nodular layer. Palaeosols, marls, mudstones and shales may be intercalated between the units. The limestones are usually dolomitised. Where they are not, any comparison made with the Trucial coast should be with the interidal and lagoonal areas rather than with the sabkha zone.

In most thick evaporite successions the evidence tends to favour a subtidal genesis for the bulk of the salts. Graded beds with basal scour structures

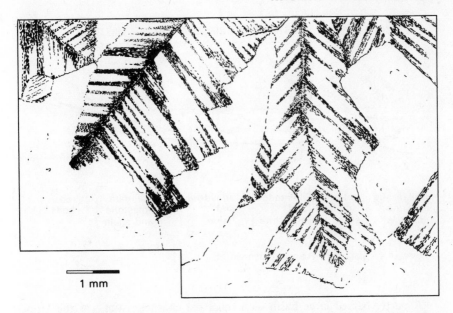

1 mm

Figure 14.3 Chevron or hopper structure in halite. The primary chevron halite appears to have been replaced by cleaner, structureless halite. Polarised light. After Wardlaw and Schwerdtner (1966).

have been recognised. Individual beds of halite are commonly up to twenty-five metres thick and collectively aggregate several hundreds of metres. The halite often shows hopper or chevron texture suggestive of primary precipitation (Fig. 14.3). Regularly laminated beds (laminites or, possibly, seasonal varvites) can be traced over tens of kilometres. The individual laminae are a few millimetres in thickness. It is unlikely that deposition took place in areas of highly irregular physiography, such as typify the margins of the Persian Gulf and Caspian Sea. Much more feasible is the proposition that the thick salt beds were deposited in the deeper parts of the basins.

In major evaporite basins characterised by large-scale cyclical successions it is assumed that marine incursions were restricted in some way, either by structural or physiographic barriers. Periodic breaching of these barriers, necessary to instigate the cyclicity, might have been instigated by tectonic, climatic or eustatic events, though the relative importance of these events is conjectural.

Minor physical barriers within basins also controlled evaporite distribution. Reefs flanking the shelf areas of the Permian Delaware Basin in Texas and New Mexico effectively reduced circulation in the back-reef lagoons, allowing the precipitation of anhydrite and halite (Fig. 14.4). Within this framework, lateral zoning resulted from the interaction of shoreward circulating salt water and diluting fresh-water streams.

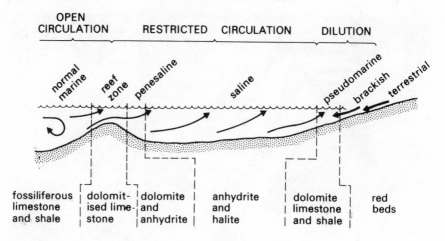

Figure 14.4 Back-reef evaporite deposition.

A partly barred large basin with relict sea characteristics was the Upper Permian Zechstein of Europe, which has an inferred extent of 250 000 square kilometres (Fig. 14.5). In north-eastern England three major evaporite beds are present in the succession, each representing a major phase of marine transgression followed by a prolonged period of desiccation and regression. The thickest bed is the lowest and it differs from the others in consisting of three subcycles of deposition, each indicating a minor transient marine transgressive episode (Fig. 14.6). The subcycles show a gradual vertical change from anhydrite at the base through to halite at the top. The sequences are complicated, however, by the presence of polyhalite, about which there used to be some dispute regarding origin. Certain authorities suggested that the polyhalite was a direct precipitate from sea-water, others remain convinced of a diagenetic replacement origin. The petrographic evidence favours the latter interpretation.

The middle evaporite bed is similar to the lower in being mainly composed of layers of anhydrite and halite with secondary polyhalite (Fig. 14.7). But the upper bed differs in showing a complete vertical cycle of salt deposition with the least soluble carbonates at the base and most soluble chlorides, sylvite and halite at the top; polyhalite is absent.

The Hanover and Thuringia districts of Germany are also characterised by salt beds which show almost complete cycles of desiccation. There are four main beds, the bottom three of which are possibly equivalent to the three British beds.

Lateral variation is a common feature of the Zechstein evaporites, as with all other extensive deposits. In northern England the main Permian evaporite beds pass gradually from a chloride-rich facies into a sulphate-rich facies and

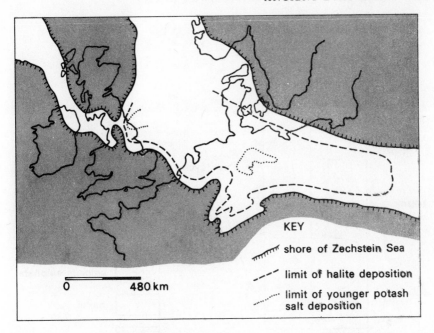

Figure 14.5 Zechstein Sea.

then into a carbonate-rich facies when traced shorewards. Red beds characterise the hinterland. Progressive shrinkage of the area covered by water and concentration of the more soluble salts in the residual waters partly accounts for this type of zoning. Density stratification of the brines was almost certainly equally important and the heavier brines probably concentrated in the deeper parts of the basin. Hence, the more soluble sodium and potassium minerals may have been deposited from heavier brine in the deeper waters contemporaneously with less soluble sulphates and carbonates from lighter brine in shallower areas.

Smaller-scale cycles involve the pairing of either halite with anhydrite or anhydrite with carbonate into varve-like layers. Where halite and anhydrite are paired the halite may represent a warmer water summer precipitate and the anhydrite the corresponding winter precipitate. This contrasts with the anhydrite–carbonate pairs where the anhydrite may be the summer precipitate and carbonate the winter. But the assumption that these pairs are always genuine varves indicating one annual climatic cycle is open to considerable doubt, especially when the thickness exceeds ten millimetres. A more credible thickness for an annual evaporite varve is 1–2 millimetres. Hence, the thicker varves must originate in other ways. Some varve-like layers have almost certainly been formed by alternating precipitation and solution pro-

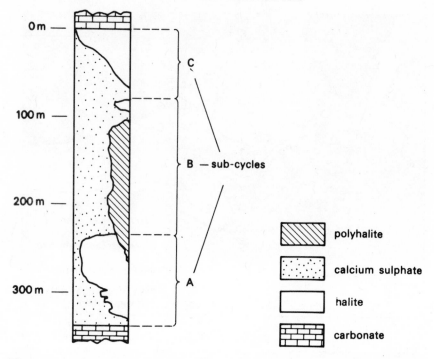

Figure 14.6 Subcycles in Permian Lower Evaporite beds. The three subcycles A, B and C were recognised in the cores from a borehole near to Whitby in north-east Yorkshire.

cesses in which a mixture of salts is subject to partial solution, leaving behind a relatively insoluble residue. For example, if a halite–anhydrite–clay–water mixture which has just been deposited is subject to changes in the constitution of the brine, then the halite may be taken back into solution, leaving behind a thin anhydrite–clay seam. The clay grade particles in these seams are usually a mixture of quartz, chlorite and illite, some of which are undoubtedly authigenic.

SECONDARY CHANGES IN EVAPORITES

Minerals probably change in composition by reacting with brines. Gypsum may alter into anhydrite as it floats downwards into denser bottom layers. Early deposited salts may react with the residual liquids circulating above them and through them and create whole new suites of minerals. And there are replacements caused by influx of more fresh or normal saline waters from external sources. The Permian potash zone minerals of north-west Europe have recrystallised on a large scale in certain areas (Fig. 14.8).

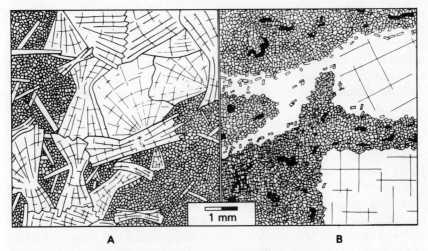

Figure 14.7 Dolomite–anhydrite–halite beds. Textures in Permian evaporite beds of north-east Yorkshire. A, coarse fibro-radiate anhydrite which has grown in a dolomite rock. B, well cleaved large halite grains set in a matrix of anhydrite with carbonaceous fragments. Polarised light.

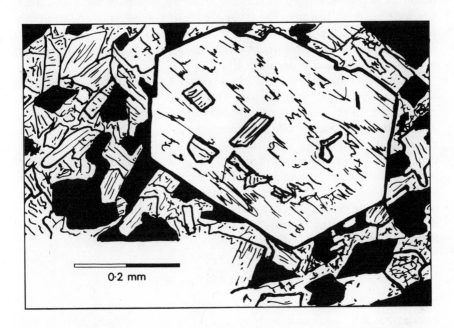

Figure 14.8 Recrystallisation in evaporites. Euhedral secondary gypsum in anhydrite bed. Relict anhydrite in gypsum. Crossed polars.

There are few salt deposits which remain unaffected by recrystallisation and metasomatic replacement. Recrystallisation textures are particularly common in beds which have been deeply buried or have been mobilised sufficiently to form intrusions, such as veins, dykes and salt domes. Gypsum and anhydrite recrystallise readily when sufficiently loaded, and will flow if the load is unevenly distributed, though higher pressures are required to initiate movement than in the case of halite. Pseudo-schistose and pseudo-gneissose textures can be formed.

Gypsum dehydrates to anhydrite at depth and the 20 per cent or so released water migrates and helps to bring about the conversion of more soluble salts. Halite–carnallite–kieserite beds may be changed to sylvite-rich beds by this process. Anhydrite very quickly hydrates to gypsum under the influence of circulating groundwaters and there are very few deposits where unaltered anhydrite beds actually outcrop. In the Delaware Basin the primary anhydrite has been converted to gypsum by percolating groundwater down to a depth of 150 metres from the surface, but partial hydration effects are known from depths of 300 metres.

BORATES

Besides their occurrence in the German Permian salt deposits, compounds of boron have been described as deposits from the waters of lakes and from hot springs, as in Tuscany. The latter are clearly of volcanic origin; and it is noticeable that the borax lakes of Western America occur in a region of recent volcanic activity. No doubt the ultimate source of boron compounds is in all cases volcanic or pneumatolytic.

In the Great Basin region of Nevada and California there are several shallow lakes or dried-up lake basins where borax occurs in association with gypsum, anhydrite, sodium carbonate and sulphate, and many other minerals. Searles Marsh, a shallow lake or swamp, occupying a hollow in the middle of an alkali plain in San Bernardino county, California, is nearly filled by deposits of sodium carbonate and sulphate, rock-salt and borax, together with mud and much volcanic sand. Here the evidence of recent vulcanicity is clear. In the Calico district of the same county a thick bed of colemanite (calcium borate) is extensively worked as a source of borax. This is a beach deposit formed on the shores of a lake. Tincal or sodium borate occurs in lakes in Tibet.

ALKALI CARBONATES

The characteristic constituent of the so-called **natron lakes** is sodium carbonate; but this compound is always accompanied by sodium chloride and sodium sulphate, together with other carbonates and sulphates. The deposits of such lakes are of a peculiar character, and often include trona,

$Na_2CO_3.NaHCO_3.2H_2O$, and gaylussite, $CaCO_3.Na_2CO_3.5H_2O$. Natron lakes, or soda lakes, as they are called by many American writers, occur in Mexico, Nevada in the United States and in association with the rift depressions of Kenya and Tanzania. Some of the lakes in the oases of western Egypt and the Sahara, and even in the Nile valley, are of similar nature. Analyses of the water of small lakes in the neighbourhood of Mt Ararat show extraordinarily high salt contents, amounting in one case to 239 parts of sodium carbonate and 53 parts of sodium sulphate in 1000 parts of water.

Extensive and thick trona and trona-halite beds are known in the Eocene Green River Formation of Wyoming. They were precipitated over a period lasting about one million years in the central parts of an inland lake undergoing rapid and repeated desiccation.

NITRATES

Important deposits of potassium nitrate (saltpetre) and sodium nitrate occur in South America. They are found in the desert region of the Pacific coast, especially in the provinces of Tarapacá and Antofagasta, in Chile. The nitrate area is several kilometres inland, and rises to a height of over one kilometre above sea level. It stretches from north to south for about 410 kilometres. It seems probable that the nitrates have been formed in small quantities over a large area by the oxidation of organic matter present in the soil, and that the resulting surface efflorescences were dissolved during occasional tropical floods and redeposited in a concentrated form from temporary lakes into which the flood waters drained. There can be little doubt that in their present form most of the South American nitrate deposits have crystallised from evaporating saline waters.

Bibliography

The papers, symposium volumes and books listed below are selected solely to introduce readers to a range of readily available literature, through which the understanding of sedimentology can be deepened. Extensive lists of references are to be found in these publications and in complementary specialist journals, such as the *Journal of Sedimentary Petrology, Sedimentology, Sedimentary Geology, Marine Geology* and the *Bulletin of the American Association of Petroleum Geologists*. Virtually every geological journal contains an occasional sedimentary petrology contribution.

CHAPTERS 1 AND 2

Bourma, A. H. 1969. *Methods for the study of sedimentary structures.* New York: John Wiley.

Carver, R. E. (ed.) 1971. *Procedures in sedimentary petrology.* New York: John Wiley.

Griffiths, J. C. 1967. *Scientific method in analysis of sediments.* New York: McGraw-Hill.

Longwell, C. R. 1957. *Sedimentary facies in geologic history.* Geol. Soc. Am. Mem. 39.

Summerson, C. H. (ed.) 1976. *Sorby on sedimentology. A collection of papers from 1851 to 1908 by Henry Clifton Sorby.* Florida: University of Miami Press.

CHAPTERS 3 AND 4

Allen, J. R. L. 1968. *Current ripples.* Amsterdam: North-Holland. 1977. *Physical processes of sedimentation.* London: George Allen & Unwin.

Brady, L. L. and H. E. Jobson 1973. *An experimental study of heavy-mineral segregation under alluvial-flow conditions.* US Geol. Surv. Prof. Paper 562–K.

Crimes, T. P. and J. C. Harper (eds) 1970. Trace Fossils. *Geol. J.* **3**, 1–547.

Davis, R. A. and R. L. Etherington (eds) 1976. *Beach and nearshore sedimentation.* Soc. Econ. Pal. and Min. USA Spec. Pubn, no. 24.

Dott, R. H. and R. H. Shaver (eds) 1974. *Modern and ancient geosynclinal sedimentation.* Soc. Econ. Pal. and Min. USA Spec. Pubn, no. 19.

Dzulynski, S. and E. K. Walton 1965. *Sedimentary features of flysch and greywackes.* Amsterdam: Elsevier.

Ginsburg, R. N. 1975. *Tidal deposits.* Berlin: Springer-Verlag.

Kelling, G. and D. J. Stanley 1976. Sedimentation in canyon, slope and base-of-slope environments. In *Marine sediment transport and environmental management.* D. J. Stanley and D. J. P. Swift (eds), 379–435. New York: John Wiley.

Kuenen, Ph. H. and C. I. Migliorini 1950. Turbidity currents as a cause of graded bedding. *J. Geol.* **58**, 91–127.

Kukal, Z. 1971. *Geology of recent sediments.* New York: Academic Press.

McKee, E. D. 1966. Dune structures. *Sedimentology* **7**, 1–69.

Meade, R. H. 1972. Transport and deposition of sediments in estuaries. In *Environmental framework of coastal plain estuaries*, B. W. Nelson (ed.), 91–120. Geol. Soc. Am. mem. 133.

Middleton, G. V. (ed.) 1965. *Primary sedimentary structures and their hydrodynamic interpretation.* Soc. Econ. Pal. and Min. USA Spec. Pubn, no. 12.

Morgan, J. P. (ed.) 1970. *Deltaic sedimentation, modern and ancient.* Soc. Econ. Pal. and Min. USA Spec. Pubn, no. 15.

Pettijohn, F. J. and P. E. Potter 1964. *Atlas and glossary of primary sedimentary structures.* Berlin: Springer-Verlag.

Potter, P. E. and F. J. Pettijohn 1963. *Paleocurrents and basin analysis.* Berlin: Springer-Verlag.

Reineck, H. E. and I. B. Singh 1973. *Depositional sedimentary environments.* Berlin: Springer-Verlag.

Shelton, J. W. 1973. *Models of sand and sandstone deposits: a methodology for determining sand genesis and trend.* Oklahoma Geol. Surv. Bull. 118.

Stanley, D. J. (ed.) 1972. *The Mediterranean Sea: a natural sedimentation laboratory.* Pennsylvania: Dowden, Hutchinson & Ross.

Swift, D. J. P. 1975. Continental shelf sedimentation. In *The geology of continental margins,* C. A. Burk and C. L. Drake (eds), 117–35. Berlin: Springer-Verlag.

CHAPTER 5

Barrell, J. 1925. Marine and terrestrial conglomerates. *Bull. Geol. Soc. Am.* **36**, 279–342.

Bluck, B. J. 1965. The sedimentary history of some Triassic conglomerates in the Vale of Glamorgan, South Wales. *Sedimentology* **4**, 225–45.

Harland, W. B., K. N. Herod and D. H. Krinsley 1966. The definition and identification of tills and tillites. *Earth-Sci. Rev.* **2**, 225–56.

Krumbein, W. C. 1941. The effects of abrasion on the size, shape and roundness of rock fragments. *J. Geol.* **49**, 482–520.

Lindsay, D. A. 1972. *Sedimentary petrology and paleocurrents of the Harebell Formation, Pinyon Conglomerate, and associated coarse clastic deposits, north-western Wyoming.* US Geol. Surv. Prof. Paper 734–B.

Nelson, C. H. and D. M. Hopkins 1972. *Sedimentary processes and distribution of particulate gold in the northern Bering Sea.* US Geol. Surv. Prof. Paper 689.

Twenhofel, W. H. 1947. The environmental significance of conglomerates. *J. Sed. Petrol.* **17**, 119–28.

CHAPTER 6

Bassett, D. A. and E. K. Walton 1960. The Hell's Mouth Grits: Cambrian greywackes in St Tudwals peninsula, North Wales. *Q. J. Geol. Soc. Lond.* **116**, 85–110.

Blatt, H. 1967. Original characteristics of clastic quartz grains. *J. Sed. Petrol.* **37**, 401–24.

Bouma, A. H. and A. Brouwer (eds) 1964. *Turbidites.* Amsterdam: Elsevier.

Chilingarian, G. V. and K. H. Wolf (eds) 1975. *Compaction of coarse-grained sediments.* Amsterdam: Elsevier.

Cummins, W. A. 1962. The greywacke problem. *Liverpool and Manchester Geol. J.* **3**, 51–72.

Dapples, E. C. 1972. Some concepts of cementation and lithification of sandstones. *Bull. Am. Ass. Petrol. Geol.* **56**, 3–25.

Dott, R. H. 1964. Wacke, graywacke and matrix – what approach to immature sandstone classification? *J. Sed. Petrol.* **34**, 625–32.

Hay, R. L. 1966. *Zeolites and zeolitic reactions in sedimentary rocks.* Geol. Soc. Am. Spec. Paper 85.

Larsen, G. and G. V. Chilingarian (eds) 1967. *Diagenesis in sediments.* Amsterdam: Elsevier.

McRae, S. G. 1972. Glauconite. *Earth Sci. Rev.* **8**, 397–440.

Pettijohn, F. J. 1963. *Chemical composition of sandstones – excluding carbonate and volcanic sands. Data of geochemistry.* US Geol. Surv. Prof. Paper 440–S.

Pettijohn, F. J., P. E. Potter and R. Siever 1973. *Sand and sandstone.* Berlin: Springer-Verlag.

Taylor, J. M. 1950. Pore-space reduction in sandstones. *Bull. Am. Ass. Petrol. Geol.* **34**, 701–16.

Trumbull, J. V. A. 1972. *Atlantic continental shelf and slope of the United States. Sand-size fraction of bottom sediments, New Jersey to Nova Scotia.* US Geol. Surv. Prof. Paper 529–K.

Van de Kemp, P. C., B. E. Leake and A. Senior 1976. The petrography and geochemistry of some Californian arkoses with application to identifying gneisses of metasedimentary origin. *J. Geol.* **84**, 195–212.

Waugh, B. 1970. Petrology, provenance and silica diagenesis of the Penrith Sandstone (Lower Permian) of northwest England. *J. Sed. Petrol.* **40**, 1226–40.

CHAPTER 7

Berger, W. H. 1975. Deep-Sea Sedimentation. In *The geology of continental margins,* C. A. Burk and C. L. Drake (eds), 213–41. Berlin: Springer-Verlag.

Grim, R. E. 1968. *Clay mineralogy.* New York: McGraw-Hill.

Hunt, C. B. 1972. *Geology of soils.* San Francisco: W. H. Freeman.

Keller, W. D. 1970. Environmental aspects of clay minerals. *J. Sed. Petrol.* **40**, 788–813.

MacKenzie, R. C. and B. D. Mitchell 1966. Clay Mineralogy. *Earth-Sci. Rev.* **2**, 47–91.

Millot, G. 1970. *Geology of clays.* Berlin: Springer-Verlag.

Moore, L. R. 1964. The microbiology, mineralogy and genesis of a tonstein. *Proc. Yorkshire Geol. Soc.* **34**, 235–92.

Patterson, S. H. 1974. *Fuller's earth and other industrial mineral resources of the Meigs-Attapulgus-Quincy District, Georgia and Florida.* US Geol. Surv. Prof. Paper 828.

Potter, P. E., N. F. Shimp and J. Witters 1963. Trace elements in marine and freshwater argillaceous sediments. *Geochim. Cosmochim. Acta* **27**, 669–94.

Rieke, H. H. and G. V. Chilingarian (eds) 1974. *Compaction of argillaceous sediments.* Amsterdam: Elsevier.

CHAPTERS 8 AND 9

Bathurst, R. G. C. 1975. *Carbonate sediments and their diagenesis.* Amsterdam: Elsevier.

Bricker, O. P. (ed.) 1971. *Carbonate cements.* Baltimore: Johns Hopkins University Press.

Chilingar, G. V., H. J. Bissell and R. W. Fairbridge 1967. *Carbonate rocks.* Amsterdam: Elsevier.

Cloud, P. E. 1962. *Environment of calcium carbonate deposition, west of Andros Island, Bahamas.* US Geol. Surv. Prof. Paper 350.

Dunham, R. J. 1962. Classification of carbonate rocks according to depositional texture. In *Classification of carbonate rocks*, W. E. Ham (ed.), 108–21. Amer. Ass. Petrol. Geol. Mem. 1.

Emery, K. O., J. I. Tracey and H. S. Ladd 1954. *Geology of Bikini and nearby atolls.* US Geol. Surv. Prof. Paper 260–A.

Folk, R. L. 1959. Practical petrographic classification of limestones. *Bull. Am. Ass. Petrol. Geol.* **43**, 1–38.

Friedman, G. M. (ed.) 1969. *Depositional environments in carbonate rocks.* Soc. Econ. Pal. and Min. Spec. Pubn, no. 14.

Hsu, K. J. and H. C. Jenkyns (eds) 1974. *Pelagic sediments: on land and under the sea.* Oxford: Blackwell.

Le Blanc, R. J. and J. G. Breeding (eds) 1957. *Regional aspects of carbonate deposition.* Soc. Econ. Pal. and Min. Spec. Pubn, no. 5.

Muller, G. and G. M. Friedman (eds) 1968. *Recent developments in carbonate sedimentology in Central Europe.* Berlin: Springer-Verlag.

Pray, L. C. and R. C. Murray (eds) 1965. *Dolomitization and limestone diagenesis.* Soc. Econ. Pal. and Min. Spec. Pubn, no. 13.

Purser, B. H. 1973. *The Persian Gulf.* Berlin: Springer-Verlag.

CHAPTER 10

Bien, G. S., D. E. Contois and W. H. Thomas 1959. The removal of soluble silica from fresh water entering the sea. In *Silica in sediments*, H. A. Ireland (ed.), 20–35. Soc. Econ. Pal. and Min. Spec. Pubn, no. 7.

Calvert, S. E. 1966. Accumulation of diatomaceous silica in the sediments of the Gulf of California. *Bull. Geol. Soc. Am.* **77**, 569–96.

Cressman, E. R. 1962. *Data of geochemistry. Non-detrital siliceous sediments.* US Geol. Surv. Prof. Paper 440–T.

Folk, R. L. and C. E. Weaver 1952. A study of the texture and composition of chert. *Am. J. Sci.* **250**, 498–510.

Grunau, H. R. 1965. Radiolarian cherts and associated rocks in space and time. *Eclogae Geol. Helv.* **58**, 157–208.

Ireland, H. A. (ed.) 1959. *Silica in sediments.* Soc. Econ. Pal. and Min. Spec. Pubn, no. 7.

Lancelot, Y. 1973. Chert and silica diagenesis in sediments from the central Pacific. In *Initial Reports of the Deep Sea Drilling Project*, **17**, 377–506. Washington: US Government Printing Office.

Siever, R. 1962. Silica solubility, 0°–200 °C and diagenesis of siliceous sediments. *J. Geol.* **61**, 127–50.

CHAPTERS 11, 12, 13 AND 14

Alling, H. L. 1947. Diagenesis of the Clinton hematite ores of New York. *Bull. Geol. Soc. Am.* **58**, 991–1018.

Bradley, W. H. 1966. Tropical lakes, copropel and oil shale. *Bull. Geol. Soc. Am.* **77**, 1333–8.

Charles, G. 1953. Sur l'origine des gîsements de phosphates de chaux sédimentaires. *19th Int. Geol. Congr. Algiers, Comptes Rendus* **11**, 163–84.

Dingle, D. V. 1974. Algulhas bank phosphorites: a review of 100 years of investigation. *Trans. Geol. Soc. Africa* **77**, 261–4.

Gross, G. A. 1972. Primary features in cherty iron-formations. *Sedimentary Geol.* **7**, 241–61.

Gulbrandsen, R. A. 1966. Chemical composition of phosphorites of the Phosphoria Formation. *Geochim. Cosmochim. Acta* **30**, 769–78.

Hsu, K. J. 1972. Origin of saline giants: a critical review after the discovery of the Mediterranean evaporite. *Earth-Sci. Rev.* **8**, 371–96.

Huber, N. K. 1958. The environmental control of sedimentary iron minerals. *Econ. Geol.* **53**, 123–40.

James, H. L. 1966. *Chemistry of the iron-rich sedimentary rocks.* US Geol. Surv. Prof. Paper 440-W.

Kinsman, D. J. J. 1969. Modes of formation, sedimentary associations and diagnostic features of shallow-water and supratidal evaporites. *Bull. Am. Ass. Petrol. Geol.* **53**, 830–40.

Kirkland, D. W. and R. Evans (eds) 1973. *Marine evaporites. Origin, diagenesis and geochemistry.* Pennsylvania: Dowden, Hutchinson and Ross.

Llewellyn, P. G. and R. Stabbins 1970. The Hathern Anhydrite Series, Lower Carboniferous, Leicestershire, England. *Trans. Inst. Min. Metall.* B **79**, 1–16.

Mattox, R. B. (ed.) 1968. *Saline deposits.* Geol. Soc. Am. Spec. Paper, no. 88.

McKelvey, V. E. 1967. *Phosphate deposits.* Bull. US Geol. Surv. 1252-D.

Murchison, D. G. 1969. Some recent advances in coal petrology. *6th Congr. Int. Strat. Geol. Carbonif.,* Comptes Rendus I, 351–68.

Murchison, D. G. and T. S. Westoll 1968. *Coal and coal-bearing strata.* Edinburgh: Oliver & Boyd.

Reeves, M. J. and T. A. K. Saadi 1971. Factors controlling the deposition of some phosphate bearing strata from Jordan. *Econ. Geol.* **66**, 451–65.

Richter-Bernburg, G. (ed.) 1972. *Geology of saline deposits.* Paris: UNESCO.

Rooney, T. P. and P. F. Kerr 1967. Mineralogic nature and origin of phosphorite, Beaufort County, North Carolina. *Bull. Geol. Soc. Am.* **78**, 731–48.

Stewart, F. H. 1963. *Marine evaporites. Data of geochemistry.* US Geol. Surv. Prof. Paper 440-Y.

Stoertz, G. E. and G. E. Ericksen 1974. *Geology of Salars in Northern Chile.* US Geol. Surv. Prof. Paper 811.

Stopes, M. C. 1919. On the four visible ingredients in banded bituminous coal. *Proc. R. Soc. Lond.* B **90**, 69–87.

Taylor, J. H. 1949. *Petrology of the Northampton sand ironstone formation.* Great Britain Geol. Surv. Mem.

Taylor, J. H. 1967. Sedimentary ores of iron and manganese and their origin. In *Sedimentary ores: ancient and modern,* H. L. James (ed.), 171–84. Leicester: University Press.

Tooms, J. S., C. P. Summerhayes and D. S. Cronan 1969. Geochemistry of marine phosphate and manganese deposits. *Oceanogr. Mar. Biol. A. Rev.* **7**, 49–100.

Trendall, A. F. 1968. Three great basins of Precambrian banded iron-formation deposition. *Bull. Geol. Soc. Am.* **79**, 1527–44.

Index